Advanced Quantum Metaphysics for beginners

By Professor Moriarty

Copyright 2015-05-22

Distributed to Amazon books and Amazon Kindle Ebooks.

Unabridged.

Edition number one.

This book is dedicated to Mr Prem Rawat for giving me the tools to know how to look within.

The magic surrounds us, it is everywhere, there is no separation from it; it is in us and of us; it is love in creation, the smile that gives hope, the open hand of friendship, it is the vibration that welcomes, the joy that makes you laugh; it is the love in the eyes of your beloved and the dog's wagging tail. It is the rain that falls and the sun that shines and the very air you breathe. It is you.

Metaphysics:

A priori speculation upon questions that are unanswerable to scientific observation, analysis or experiment.

Excessively subtle or recondite reasoning.

The philosophical study of being and knowing.

Essential nature or underlying reality.

THE MISSING LINK

The quantum theory principal can be applied to anything that has a nature it can be applied to notwithstanding anti quark expressions, solar anti masses, holy moly holy grails, infirm grandmothers and children who shout too much in so much as c squared will allow in amongst the roses in bloom of any conducive theory yet to be put forward.

Nevertheless, never let it be said that time doesn't have some kind of influence in the hereafter of the here and after and if my calculations are correct this may be so unless I write in long-hand and then it takes much longer; therefore relativity theory may be a better option and mosquitoes aside this may be the case.

However, all things considered and in their proper place we cannot allude to such without taking into consideration professor Gustav's hypothesis when he looks through his telescope, and that is that when all things are considered it all may be too much to find any real clarity on the subject at hand, so much so that we could be here doing our sums for a very long time indeed.

Furthermore, as Sherlock Holmes used to say, the quality of the quantitative easement, as has been mentioned before can only bring more questions to bear on the problem already over-burdened in the Quaaludes of the strained imagination in free flow which may be an oxymoron but let it be so for now for want of a better exclamation, and dare I say: eureka would be nice but until that moment comes the lumbering foundations that are present will have to make do until of course one can blast off without a backward glance or a thought lost in query; but not everything falls in the face of gravity.

So until someone comes up with a better theory, in that time to the umpteenth degree without any extraneous thought wasted, the quantum theory principal can be applied to move forward any idiosyncrasies hung up to dry for want of a worthwhile and heartfelt reason to pursue them any more than has been tried down through the ages of mankind.

Which brings me to my second point in this: subjugates of entelechy notwithstanding that bounce around and collide with the hadrons that are not there for all intents and purposes in any discussion under the sun, which leaves us where we began so long ago; and so saying we shall now move on to the next item on the agenda: the missing link.

The missing link:

Despair is an odd paradox of the monkey mind that has no use other than fear and repression where the outcome is usually depression.

Watching for a favourable wind would be beneficial and although there really isn't any time, take a clock with you just in case you meet the missing link and have nothing else to say but: on your marks, get set, go, and then sit back as the missing link does the circuit and all you have to do is time it.

Quantum to popular theory we cannot advance any further without going back to our roots. But roots being roots and not telephone numbers we may find a marble of turpitude squatting there, because as you know, a vacuum usually wants to be filled, and because the grass roots of a thing may not be more than its sum of parts this may be so.

The familiar on the other hand has far more import and should be treated carefully in the management centres of the word; and if you haven't got so far yet then look up numbers of abbreviation in the cowpat theory by Terry Mac Adams in his book: the worm theory of evolutionary relativism.

"Between you and me and what we never say I would say there's a huge space that we try to fill by what we do say," said the elastic band saying what it felt to be so.

"You don't say," said the rubber onion, breathing in and out hugely.

"Yes I do say and I have a feeling that if we said less we'd say more," exclaimed the elastic band, saying more to say less.

"You're going to have to wake up soon you know," mouthed the rubber onion saying not a lot.

The points of no return:

At the points of no return where quantum theory can no longer go with any accuracy, where in the end it is what it is, theory breaks down to where it can be said that there are ghosts waiting to get in and that the broom cupboard is no longer a safe place to hide.

This is causing no little amount of consternation among the scientific community and as you can imagine there is some hair pulling as well, said scientists not being as rough and tumble as they used to be.

Dr Who, who shall remain nameless has gone so far as to say more than can be quantified in such a small book; needless to say, this is being looked into by the experts, whoever they are; and where they come from is beyond deductive reasoning; suffice to say, we are all here one minute and gone the next, just winked out, like a quark in the soup.

So if you do find yourself up near to the points of no return, be it doing the Thursday freebie on Friday, or head in the clouds, or anything else in between do remember to switch off your headlights or your battery will go flat.

And so, that's the secret of the universe; but it won't do you much good unless you can fly.

On the other hand it's possible that the void is filled with particles so small they can't be detected and this is why light gains a top speed of 186000 miles per second and goes no faster.

In a true void light would not be held back; but then without these extremely small particles light may not be able to travel at all.

For all intents and purposes a void is a void if it is deemed empty of all measurable substances; but it is possible that there is no true void and that it is filled with a substance that is no substance and so can't be detected or measured and that is why light travels in waves and particles both.

Sound won't travel in a void because the particles are too small to carry the sound waves; but light, being a higher energy makes use of this invisible substance to travel along similar to the way sound uses the atmosphere to travel.

This has yet to be proved and so remains a theory and like Zen this can best be understood when you don't say anything about it at all.

NON-REALITY DUALITY

Way down below in the kitchenous sinkhole of materialism of mortality where the soup of stuff was being stirred by the loving hands of the soup maids was a spiral of discontent along the curved space time of another reality (there being so many of them it is sometimes hard to keep up with it all) where the seven spiders of oblivion were having a discourse on what could not be said and it is around here that butterfly Joe the butterfly came along and beat his wings causing an effect, not of one but two singular exceptions in the space time continuum that made another reality open up and this is where anyone with any sense to do so would go and get a cup of coffee and forget all about quantum theory and perhaps even put the cat out.

Quantum theory aside for a moment; when you come across the kundalini borderline as you will do sooner or later, don't be afraid, it's just another singularity plural around a big seeming fire that you will have to dive into to progress on the road of knowledge.

You could liken it to a black hole if you want to but for me it is where everything is stripped away as it has to be so that all that is left is the being, that singular nothing that is left when all things have been discarded and there is nothing left but the baseline of who you really are and then in that state your quest becomes fulfilled as up from the depths draws near as near that which cannot be divided or separated.

This is the holy grail of quantum physics; no machine yet made can measure this or record it and no words can describe it and so perhaps quantum physics has gone as far as it can go in its present form when this knowing has been reached.

It is feasible that a time will come when quantum physics will merge with other disciplines such as Zen Buddhism, Tao practice or Hatha yoga to finally understand that there is no final solution and that life is ongoing and flows its own way regardless, perhaps to the conclusion that the mind is enough to penetrate the mysteries up to the point where only the heart can go further and that the answer is where it has been all along in the heart of the human being and that although you can write seemingly forever on endless paper it can never be explained or defined in this way, or any other way, for the very defining of it is like a beam from the sun landing on Earth and turning to dust and then put in a bottle which is then displayed in a museum with the legend written on it: this is what the sun is.

For the more advanced and enlightened quantum theorists the turning point of the understanding will come in a blazing epiphany or even perhaps just a turning in the intelligence layers where an understanding will come and all things will fit into place.

This is the first step to knowing; from this knowing all things can be penetrated; for those in the know this is true enlightenment and it is here where all traces of imagination fall away and the naked being can be and all concepts and theories at this point are left behind for knowing supersedes everything.

The time is nearing when no other mechanism but the enlightened human being will be enough to advance in the mysteries and to enlighten others, and in this way, when the student calls the master will answer.

The old paradigms are no longer serving humanity and so are falling away into the illusion they came from, although some still use them as stepping stones out of a lingering fear but that also is disappearing to be replaced with a courage to understand the deeper mysteries and in this way the student, be it in quantum physics or just layman thirst to know will begin the inner journey to what has never been closer and yet has always seemed so far away.

This far awayness may have been a deliberate and perhaps unconscious miss-direction by institutions that should know better, but the great awakening of humanity that is now happening is beginning to question the miss-directions and so come to know for themselves rather than believe what has been told down to them for so long a time in our history.

The reason it is happening now is because when enough people know a thing it seems by some unknown design to spread to the wider populace by a remote osmosis and so the truth spreads and is unstoppable as has been proved in many ways, such as when a group of monkeys on a remote island learned how to use a tool, and then other monkeys far away on the mainland also learned how to use this tool, with no contact between them.

The knowing had spread through some invisible means off the island to the larger community.

Quantum physics has come to the point of understanding that the emptiness between all apparent things is not so empty after all and that it is perhaps real and it is the apparent things that are an illusion and that solids are only solid because of the speed that particles spin at which make it seem solid when in fact it is mostly empty and yet full of something that has not yet been measured or seen.

This makes some people hop up and down in frustration to find all their endeavours have brought them back to where they began to begin to understand; and perhaps new theories will be made to understand ever deeper and yet always to be brought back to begin to understand.

Such a quandary, that the tools made are of the illusion and can only measure the illusion, for what is real can only be felt in the heart.

QUANTUM THOUGHT

The truth, like an important rainstorm has much to say in its nature and is usually profound, but not always, it can fizzle out or even fade into the next knowledge base to come along; but in so much as it's raining then the newness of that is its reality.

Some perceived truths are a waste of time in the time and non-time of it all and quite often lead down wrong avenues, as the Greeks with their geometry patterns theory that led the minds of the time to bully up their ideas for one and all that spread far and wide and although the Greek theory of geometry has its place and can be used as a stepping stone, recent quantum theory has surpassed that in the search for the basic building blocks of the stuff of all that is perceived that is; but so far no solid platform to build from has been formulated beyond theory.

But as has been said, it may be time to go beyond theory to find a solid foundation to build theories of knowing from.

Explanations abound and could fill the milky way with their theories, but a theory is a theory is a theory and has its basis in what if and then calculate up to prove it in some way until the next theory comes along to either disprove it or show the next stepping stone.

"Oh please," said the little well to the ocean, "what are you trying to tell me?"

The ocean raised its eyes to the Heavens and got out the old fallback blackboard.

"Now pay attention," it said and began chalking squiggles over its surface.

"...and so, E equals energy, M equals mass and c is the speed of light squared which is something Einstein thought up a while back which made perfect sense at the time but is now looking a tad outdated, what with one thing thing and another that has come along since then."

"Oh," said the little well squirming around and wanting to go out and play.

"Attention deficit disorder," screamed the attention police and jabbed the little well full of drugs of conformity to make it behave within the parameters of all the many laws of learning in the established establishments establishing the set education so it would fit in to the system set up to perpetuate the system for the elite to remain at the top of their pile of wealth.

"And now that we have your attention, what's left of it, repeat after me: one and one equals two."

"Duh," said the little well.

"Ok, this one's ready, ship it away to the pits, type D, class A fodder," said the voice of the system and off went the little well without a sound to quietly die a living death for the rest of its existence breaking rocks for the system while all the while the elite lavish in their luxury with more than enough to last them for ten thousand years each.

Which brings me to my next train of thought: why is the speed of light set at 186000 miles per second, and is there anything faster?

As to the first one, no one knows why but that it is a fixed constant and can be used to measure other things by.

And, nothing is faster than light.

If time and space are not real as proposed by the Eastern mystics and the new quantum physics thinking then that which is indefinable is everywhere and any part of it can appear instantly in any other part or field of view by the observer of it in that moment which makes it pretty fast and certainly far faster than light; and if space and time are illusion yet set by immutable laws where light as an open constant yet fixed at 761 million miles an hour then the mind boggles at how huge the known universe it.

Running backwards at the speed of light would be a pretty neat trick but the legs would wear out immediately and the rest of the body would disintegrate in the atmosphere; but of course I hear you say, light only reaches light speed in a vacuum so it would not be possible to go that fast while still on planet Earth.

How long does it take light to reach light speed?

If light was sealed inside a box would it disappear, and if it does, where does it go?

Suppose a spark of light could be generated inside of a perfect sphere that was a mirror on the inside, would the light bounce around forever or would it extinguish? Or to put it another way: if the sun was to extinguish instantly would all the light also disappear?

Perhaps the answer lies in the stars of those suns that are no longer there and yet the light of them still shines as if they were still there.

Einstein proved that light is its own form and that once made it travels on regardless of source until it becomes what it isn't.

So, what is light? Light is energy and has a momentum of photons in waves with particles here and there dancing the dance; but relative spatial and temporal specifications aside and words of common usage of course, light seen from a philosophical point of view is good for the observer to be able to see.

Reality transcends ordinary language and advanced logic is bound within its concepts, in other words, if logic is a space-ship in outer space the occupants inside are bound by its life giving embrace of atmosphere and walls and although it may be able to travel around a little bit here and there, still, everything known must be known from inside that sphere of holding.

So, if I'm here and not here at the same time then where am I really?

To answer that question we must digress beyond the point of no return where we return again and again and again in the age old karmic cycles that sooner or later we must break out of to go deeper into that being, non-being that has no name.

This delicious circle of duality has its hind feet in the gravy of the law until the lesson has been learnt and then we are free to fly away.

The such-ness of now as shown by the dancing god Shiva to the Hindu followers to name one god is of such perfection no words can describe it; similarly the end game of quantum physics holds much the same outcome relatively speaking and so language has to be gone beyond; but we're a little bit stuck on the page so far as that is concerned so let's do an experiment and throw a stone into the mind's waters and see what the ripples do:

In a clean pond of thought the stone might become absorbed and transmuted thereby becoming what it is surrounded by; but the opposite is not always true.

A muddied mind will take on the energy of the stone and may even fly off and become a storm after which the ripples can be detected for a long time afterwards and quite often go beyond the borders of its boundaries and affect other minds.

Transmuting the stone into mirth can have profound and immediate measurable effects upon the ripples whereby influencing them into another form of energy that can have a different affect to the original stone.

This effect is called balancing the waves and is transmutational energy caused by thought in its basic state.

The power of thought is able to influence matter through measurable electromagnetic energy generated within itself by thought alone.

This phenomenon has baffled scientists since it was first discovered and remains a mystery to logical thought to this day.

A still mind on the other hand has no energy of cause and so no effect is generated, whereas a muddied mind makes causes left right and centre and consequently the effects can be measured and even seen by the naked eye.

So in the such-ness of now with an untroubled mind the dance can be enjoyed with no fear of cause and effect.

This frequency is of such a high energy only the human being is able to experience it and indeed no other device is allowed in. Which brings me to my next point of reference, that the outsider has to go through the door to experience this, and to do that all made devices have to be left outside, so it is small wonder quantum theory and indeed all other disciplines fall short of discovering the holy grail they try so hard to discover, for to define it is to make it into what it is not; and undefined it will remain a mystery.

This quandary is called the tuppence worth of buddence effect and is not to be confused with basic quantum theory or any other theory for that matter.

Now, out past the edges of all things where nothing is perceived and light has yet to go, the boot strap theory might not have much effect and would maybe not do much, and in fact, in such a vacuum might even shrivel up and disappear altogether, but nobody knows for sure and what does it matter anyway, if there's no one there to see it then who cares? And such an experiment may not be of much use except perhaps in a purely distant academic form of research that could never be carried out unless the aliens had a super fast space-ship to take scientists there and back, but then it would no longer be empty space and so would negate any resulting research made by being there.

So how do you observe without observing?

It seems thought alone is not enough.

"Come in CQ; Come in CQ."

"CQ calling, what is it you want?"

"I need an answer to an insolvable problem."

"Oh-oh, CQ over and out."

Not all theory can be explained, quantified, qualified, or calculated and so remains a theory until new thought produces an answer but by that time yet more theories will come along that have to be proved, so in a way quantum metaphysics is a waiting game that needs lots of patience.

A SMALL PARTICLE OF INFORMATION

The common sense principal must surely apply to the new quarky on the block that scientists have called a pentaquark, which is a new class of sub-atomic particles; and if not then the quarkus effect will apply.

But the full-moon cannot divide us unless we step into the oracles in between and then only when we're calculating in the string theory mode, and if such said calculations were to be correct then we have so far got to here without too much of a probability cause to fire the neurons beyond anything they cannot harness to go further than what is too far; too far being what the senses can extrapolate from any known or unknown, depending upon which side of the point of any new discovery you may be on.

If we let the string theory fly in the face of all opposition we get Faraday's exception sailing into the sunset daisy where most beliefs are born which may not be the straight response we might look for but for now it's a step in the right direction and which does not stretch common sense beyond all reason.

'What is the answer?' is a malady we have burnt out on too many times; and so 'who is the answer' may be closer to the truth than all else we have trusted to date; if all things have an answer then it may be closer to home than we have so far gathered in our perceptions that have run true or otherwise depending on the point of view we are using in any moment.

Suffice to say, the answer we would have is a fast character we have spent too much time chasing, and sometimes regardless of the consequences we have barraged the storm with our teeth chomping the bit to find answers with the supposedly safe house of our expectations and perhaps closer to tomorrow than we would like.

Parallel repetitions aside, we cannot assume we are anywhere but where we are that comes to us out of the blue in any collider we can make, and reportedly, if tomorrow didn't come by design we'd all be in trouble; and as the aliens are fond of saying: "we are not the first to land here from far, but this is our home now."

Moving within this plane of existence requires exceptional thought processes that sometimes have to come seemingly from out of nowhere to build a bridge between the separation we find ourselves in of knowing and not knowing.

If we'd had the internet in the long, long ago then we'd know all this and more well by now and the narrow confines of ignorance would have no power to influence us, for knowledge is most powerful against it.

So if finally we are going beyond the bounds of narrow minded thinking that most usually forces itself out of its box to sway the mainstream consciousness to its reality then this could be an exciting time for all of us and we may find ever smaller quarks that lead us sooner or later to the discovery that we are all one and that there is no separation, assuming of course that we can assume such from a particle of information we can't see with our naked eyes because it's so small and may not exist except in our dreaming.

This of course may be no more than the distractions of some spineless dog that comes to lie with any stranger on a bed of ice that freezes us into ever smaller discoveries that lead nowhere.

Which path will we choose when it comes right down to it: do we excite the chattering monkeys to ever greater heights of noise, or do we silence our minds in the knowledge or our knowing?

However, when you sink to the bottom of the ocean of thought of all this it won't matter if it's raining or not, and like a dog that shakes itself while still in the river, life will carry on.

There again, if you find yourself landing on Mars without a helmet then write home for them to send one quick, and hold your breath.

If on the other hand you find yourself grinning from ear to ear then you've gone as far as needs be and it's a good day.

THE EXPERIMENT EVENT

The Hadron Miximus Generator frequency machine:

Over in the other fields there is some thought that something peculiar may be happening and that the world may end sooner than predicted.

This of course is outside of the spirit of things, and regardless, the Miximus generator will be fired up again momentarily as soon as we've cleared out all the mysterious pieces suddenly found there after the last spin-up.

Particle colliding inside the Miximus generator is theorized to be as safe as anywhere else it can happen and if the experiment event is shut down we may never know what will happen when two particular particles collide and what they will produce.

So far we have discovered that not every pretty face can hold a smile in close proximity to the Miximus field at thirty percent power, and we are still determining why hair stands up straight and turns blue, and other effects still to be analysed.

Update:

The latest news from doctors far removed from the effects of the Hadron Miximus generator have yet to say why brain waves become mixed up in anyone inside the field.

This has been diagnosed from examining the few individuals who have so far wandered away from the effects of the machine that self generates its own power and cannot be shut down by outside interference.

While all contact has been lost from those sent to investigate, it remains hopeful that one day soon a way will be found to shut down the machine.

Another effect that has been noted is that all individuals wandering out of the field also have lost eighty percent of their intelligence and are now to be considered idiots and as to whether this is permanent or not; no one can yet determine and so the prognosis remains indeterminate.

Although the latest news states that the field is steadily growing outwards and that people are being moved further away to a safe distance this has yet to be verified and so cannot be said as true.

It can be confirmed that the government have been informed but are not alarmed and have elected to remain on holiday far away.

My latest theory which I will call the Moriarty Miximus theory states that the field feeds itself on intelligence and that it now has gone too far and cannot be stopped and will cover the entire planet in days leaving the population as idiots and what will happen then is anyone's guess.

I surmise that the field will digress back to its origin taking with it the entire intelligence of the planet to lie dormant until signs of increasing intelligence can be perceived whereby it will advance once again to feed.

This may mean that there will be no new discoveries in the quantum field or any other field for quite some time and possibly until the machine has fallen into disrepair in around a thousand years or so from now.

This is why I am busy trying to record as much as I can so that if the population of the planet ever becomes intelligent again there will be a record of advanced quantum physics and what has happened in the Miximus field for future generations in the distant future to carry on the experiment and so that beginners can start again from where we left off before the great sleep stopped it all.

The field is reaching even to here on the far side of the planet where I've been flown too to find a way to stop it but even here, the field although weak begins to affect me and my calculations in any given moment, but my IQ is still above normal and so I hope there will be time to finish what I have started before all reason leaves me and I become the idiot as so many so far have fallen prey to.

A VALID QUESTION

In the advantages of having a clear mind all things can be known, but not always at the same time.

So in the mass consciousness where an idea can spread like a virus so much so that when the idea comes up: 'what are we doing here?' everybody goes: "huh?" But it's a valid question, what are we doing here?

For most people it is to work, make money, vacation if they can afford it, have a few kids, grow old and retire and then pass away like some worn out cog in the machine that has many replacement cogs waiting in line.

Quantum physics can quantify all this to the nth degree, but then what?

Sometimes there is too much to see and in the looking is found the part but not the whole.

It is known that the part is in the whole and to know the whole you have to see all of the parts but the parts are too small to be seen all at one time, so to know the whole by the sum of its parts will need another way than current scientific science uses. It is of course a question of knowing.

CROSSROADS

We are at a crossroads where exciting things are happening and the static quandaries that have beset us are falling away to leave only the quantum questions to be forever answered and all this has happened in the time it takes for a tree to grow.

Developments come thick and fast and are collected by the eager beavers who run around at light speed with pens that never run out of ink. (Maybe because they are never used and are left in the pockets as fashion accessories; whatever; that's not important.)

However, there is a certain expanding in some quarters towards the old ways where organic and natural tendencies served humanity for thousands of years before modern science came along and bulldozed all that into obscurity by blind thinking and forced ideologies that are now being seen for what they are as pie in the sky falsehoods.

So much has been lost to humanity's collective knowing during the dark times that it is a wonder we all didn't digress fully into the darkness and pass from existence.

...It's all a game and like a chess piece on the board you'll go where they want you to go; but everything they've taught you is a lie or a half truth and until you open your eyes you won't see this. So until then you will be seriously following their rules to fulfil their agenda while believing all the time that you have a say that is worth something; but your worth to them is a passing moment to build their empire...

Finding the truth is now the priority of human kind.

EXTRAPOLATING AN OXYMORON

"Frequency, energy and vibration," as Nicola Tesla said, and, "all is energy and match the vibration," said Einstein, and Max Planck discovered 'energy packets,' but Einstein called them 'quanta,' which led to quantum theory.

If the hadron bootstrap theory was enlarged exponentially would it ultimately show the universe repeatedly making forms that are self aware which as a logical extreme shows the universe being aware of itself?

The Leibnizian monads came from a fundamentalist way of thinking in which he said: "they were forces imprinted by divine decree and are the elements of all things."

This view is in opposition to the bootstrap theory and the view of Eastern mystics that have the view that space and time are interpenetrating, which could be seen as: 'everything is everywhere at every moment.'

So, are there varying levels of consciousness, layers perhaps that consciousness penetrates to attain pure enlightenment?

And is there such a thing as one hundred percent consciousness?

Levels of higher intelligence may not be able to answer this without theories.

In the Eastern view and in the view of modern theories everything is connected and no part is fundamental.

So to extrapolate all this into its quintessential essence of probable and inescapable theorem for the intellect: nothing is what it seems where you look further than what seems to be so before you look.

Nutshells being the in thing these days so I may postulate: 'beyond here (here being any point of reference to go from) lies only what can be seen when first seen before beyond here came into being, meaning: 'time and space being an illusion and not real everything is a construct of the imagination and what is really real is beyond the imagination and paradoxes aside where oxymorons abound, the fundamental truth of the universe is where it has always been to be found, and so looking away from that brings only more things to look at.

Angels may send coincidences and white butterflies to say they agree with this and some may see this as a by the by kind of phenomenon that can only be explained from a spiritual point of view and frame of reference, and of course there are widely differing views on this to the point that most quantum scientific scientists succumb to hair pulling and some may even go so far as to wail like stricken ghosts in their forever tombs of wailing beset by the dust of idiots who have no frame of reference to hold true to and so drift about in exile extrapolating facts as they come into awareness and then shedding same as they move on to other points of view in the void of their unknowing; and it is this state of affairs that seems to be peculiar to those humans who are removed from their true nature.

All things know their nature as being a part of them. Turtles lay eggs in the sand and then hatch to swim in the sea and then eventually come back to the sand to lay more eggs and in this way perpetuate their nature.

Whereas humans find other things to do outside of their nature and so duality is formed; duality being a concept confined within the minds of human beings and some would predicate that this is the lot of being consciously self aware.

And herein lies the dichotomy where we may extrapolate further using self awareness as our tool in the verisimilitude of the flowing mind to go beyond what we now know and explore in the new paradigm thinking where process outmodes structure and so the part is the whole and the whole is the part; but where do you differentiate between them?

This is an oxymoron and needs other thinking to understand in part and whole what is going on.

So relatively speaking, atoms and such like parts can be broken down into smaller bits which can leave you thinking a long way from the whole; but the whole is full of space, which is another oxymoron, for how could a solid be full of space?

Seeming forms vibrate so fast around the seeming solid that the solid appears solid, and touched is resisted by its energy until a greater energy is applied whereby breaking apart the whole to become parts; this applies all the way down to the subatomic level and at this level, protons that have been disrupted yet remain protons but the whole disintegrates into parts.

The tide of expression that falls where we least expect a return on our giving and yet all we give is in the falling where we fall to give; so give with no expecting of return to light where we fall into any moment of giving...

Does a subatomic particle have feelings?

If so, this would imply that the whole is made up of the part and that all parts are an intrinsic model of the whole.

To have a positive inclination towards that aspect of modular intuitive mechanics of quantum theory eventually would leave one exhausted to be so split up from the whole in such a way that one derives little satisfaction in such duality and only in coming to rest between the doing and the non-doing would one come to find relief.

Turpitude does not relieve boredom but rather enhances it and this is why extraneous belief cannot jump over the moon in a month of Sundays faster than knowing, in this universe anyway.

So is thought a barrier to higher thinking?

In this way the secular ideology of words must be relegated to the fifth occupation of the cause of limitations which we haven't got to yet but will do so in the coming chapters but you can hang onto your seat if you so wish and even eat popcorn; and so with your tongue in your cheek we shall proceed.

ALGORITHMS OF ALLEGORY

In the algorithms of allegory the residual occupancy of a stray thought leaves behind some form of its passing and imprints the functions of its properties into the ether.

And in the post pre-natal tensions of a new idea we must batten down the hatches and infuse the focus with direction to not let it slip away.

And when two opposing ideas meet they begin stripping each other of information, much like Hancock's vampire theory from his book: the long word home.

In the emerging hyperbole where this exchange takes place, a tit for tat kind of thing where subconsciously things go on in clandestine meetings with the conscious mind being blown in this wind and trying its best to catch some of it to make such a brilliant exposition of new age thought, a means of recoding it all as it comes would be really something, if only someone would invent some way of doing it.

Now that we've touched upon hyperbolic elevators and have lost touch with that direction for the moment and oh where would we be without the flux capacitor generator to cool our ideas down to be useful if a little fumbled from one hand to the other.

Now then, if energy is all around us and regardless of whether it can be created or destroyed for the moment wouldn't it be so good if we could harness it, free energy for all?

Maybe a bit like putting some in a box and boiling water for tea and then opening the box and letting it go on its way again, much like a tramp that comes for breakfast and eats all of your donuts and then tramps on again somewhere.

But how to make use of it in a useful way?

Oil makes fuel for machines but is in finite supply. Water on the other hand is plentiful and machines are now being made that can extract the hydrogen and use that as fuel. The hydrogen atom consists of just one proton and one electron: one plus one equals two.

Neat.

There are many forms of energy: wind, solar, gravity, wave power, electrical, chemical and many more.

As you know, atoms have a structure of nuclei and electrons and they in turn consist of protons and neutrons or nucleons, and so on, and all this is in the form of dynamic energy and indeed it has been shown that the activity of matter is the very essence of its being.

And so in the onward march of humanity we come to understand more and more ways to use this energy.

GHOSTS IN THE SUBFUSION ANOMALY

"How am I?

I am most often than not what I seem, but I am not always how I seem and that is how I am unless I am something else; to some I have been all things and to others I have been less than nothing.

In between these two is a compromise that's fine and ok and of the passing moment that is either here or there but better here and now.

But all in all I'm so glad you didn't ask me why I am," I said breathless all of a sudden for no reason.

"Oh what are we talking about now?" asked my wife yawning.

"When the strange urge comes do we not shiver, and don't we fall down the cliff of despair when we fall down it, and is there any mountain we cannot climb?

Wrap up the answer in a painting and leave it in the hot sun and come back later so see how it is.

And ah yes, Gurgling Pete, I remember him well, the one who fell out of his tree into the ocean and was never seen again; and Mandy Marbles, who was a witch with blue hair and married God and moved to a desert island; and they live there still, swept away in the immediacy of their love and yearning," I said for want of something better to say.

"Crunchy Chinese horse chestnuts," said my wife in reply.

"In the indeterminacy of language the Sapir-Whorf hypothesis proposes that languages are tied to their communities; what then the lone gypsy oh?

And Wilhelm Reich on the function of the orgasm got put into prison forever for saying stuff..." I said, saying far too much.

"As opposed to what?" asked my wife picking bean sprouts out of her teeth.

"All mental giants have to be careful when they stand up in case they bang their heads on the low ceiling of all the old paradigms that are held in place by the ones who won't see beyond them," I said.

"So can you drown without an anchor or skip without a hop," asked my wife looking at me closely now.

"When the arrow of love passes through your shield into your heart you'll know, but until then the sands of time will take you away further than you can return, until you return," I said.

Momentous then the decision to wake up and turn off the radio of all this," said my wife hopefully.

"Who will be my friend in the friendless place, and who will help me with my weight when it feels too heavy, and who will pick me up when I fall and whisper secret nothings through the wall?" I said wistfully.

"Well the indeterminacy of language of course," said my wife, yawning.

"When you take a little faith and stick it in the jackpot machine you get thin soup forever; and now it is that senseless sneezing time again," I said and began to sneeze.

"Keep it simple," said my wife and went to sleep.

Grandmother inertia was in her rocking chair soundly snoring when the ghost of an old idea slipped in through the back door and silently passed her hoping she wouldn't wake. The cat on her lap, awake now, stared at the ghost as it silently crept by, and the soundless scream that came through the wall disturbed no one and so dropped into one of the buckets of darkness stacked high to the ceiling and was heard no more.

The clock in the passageway of the stairs thronged its hourly chant and sent the ghost up to where the dreaming was where the chamber of its passing accepted it to slip away and the long day was done.

But my day was just beginning for the hot paper of a brilliant hypothesis won't move a mountain until it is put into practice.

Now that I've addressed the uncertainty principal in Quantum metaphysics we shall go on to:

The quest for reality:

An old door, pretending to be one fifth of an illusion wanted to put in a good word and so began banging its drum...But Mr Levinson the oriental had something to say quick-like as he jumped up and down on the hot ground, his burnt feet smoking..."Oyo-ah, oow-ah," he went and then danced off to find water and shade.

I was left nonplussed and without an explanation.

The old door began making inroads and would not take no for an answer even though it was such a simple word and so began a monologue of its diatribe to say what it would talk-speak to converse all it could upon my clean white page that truth be told had ideas of grandeur and pots of gold.

The watershed of all this was an old question that later would marry up to the old door, if I let it go on; but the old door had already downed three cups of coffee and so was being carried away with it all to the mark of no return which was all well and good with me, because, frankly, what can an old door possibly say other than to squeak or bang in the frame?

"And therefore to think, dear fellow, that that is all I am because you think it so," said the door grinning hugely from ear to ear.

'Such strange goings on,' I thought, taken aback once again and so soon after the last time.

"Come, open me and see what is revealed," said the door not at all shyly and still grinning hugely.

Thoughts came to say that if I was caught talking to doors I may be thought of as eccentric, and so I refrained from answering.

"To much peer pressure I see. Ok then, by the numbers, hup..."

"Whatever you say it is or isn't has nothing to do with what it is, and in fact the intellect should be put aside in favour of the heart," said the door and began to dance.

'There have been many humungous ideas that have flashed away down the pan before I could wrestle them onto the page,' I thought I thought in time.

Are we to dedicate our meagre rations then to the mouth that calls the loudest, and for this we live our life?

"Oh by gum it's me tooth."

"Ah me hearties, it's his tooth, his tooth."

"And what does ee do with his tooth?"

"I eats me dinner with it I do, I do."

"Pray tell, how so, we want to know."

"I digs it in n gums it down and charges a sixpence to half a crown."

"So it's his tooth, his tooth that makes him rich."

"Aye it's me tooth, me tooth, and tis grand it is."

Chorus:

It's his tooth, his tooth it is, it is, it's his tooth.

It's me tooth, me tooth it is.

Oh it's his tooth it is, it is and that's what makes him rich...

"Yoo hoo."

"Let's gummy im up like between us, umm?"

"Ee's mine, I got de tooth."

"Not fair, how come you got a tooth?"

"Lucky I guess; nik, nik, nik, nik."

In the dreams where I hide I peek out from the illusion; the bottomless pit glares back at me full of ghosts swirling in a mist. All is unreal, so I go back to sleep.

Hauling up the moon:

Moonstop and Beamstrap were late again pulling the moon up into the sky, a pair of numbskulls to be sure.

But the puzzle daily becomes more absorbent.

There was once a time when coal holes were built into the pavement to pour the coal down into the coal cellar. At the time this seemed a very civilized way of getting the black stuff into the house; and then one day all the coalmines shut down and everyone changed to using clean gas, so they said; but some still use the coal holes to get coal in.

Similarly, when rubbish was thrown in a black-hole it disappeared never to be seen again, it was thought. But that is not the case, your karma will always come back to bite you in some way for all is connected and nothing disappears forever.

A loophole on the other hand is a different kettle of fish. Going down a loophole will take you to an Alice in Wonderland kind of experience, but don't worry, it's not real, but there again, nothing is real so there's never ever anything to be worried about, although sometimes you may have to endure, but that's another story for the long dark night of the soul to be told around the camp fire of the vanities where the ego burns to a cinder.

So anyway, one fifth of an illusion is not really very much at all and in fact may be less than nothing which in the time of the Greeks was really not a lot at all. But these days with the current thought that nothing may be filled with rather a lot then nothing may be more than it seems; but an illusion is always an illusion.

So it is possible that having nothing may mean you've got a lot; so if you've got a lot of nothing then you may have an awful lot.

Something, because it is in space and time, and space and time being an illusion means you have an amount of illusion which means it is not there; so following along this line of logic it is better to have nothing which is a lot more than something which is illusion.

Perpendicular to this tail-end subtraction, the derivative of this could be the beginning of an exciting new thought-stream of theory that can be likened to looking in a mirror and seeing something that's not there.

Living in this mirror would be colourful in the extreme where no two moments would be the same or normal.

Normal? How could anything ever be normal when not a moment is the same as any other? Normal is an illusion and though similarities abound, each moment comes as unique. So, if everything is a unique event then how amazing is this one moment in existence...

Formidable forebodings to the contrary this is not so hard to achieve and in fact takes no effort at all.

It is the unlearning that is the effort and so, depending on where you are in the evolutionary queue will depend on how much effort you have to put in for it to be effortless.

Treading on water for instance can seem effortless from a layman's point of view but the physics behind this seemingly impossible and weightless feat is far beyond most mortals at this time and yet in the animal kingdom it happens all the time and is possible from a standpoint within their nature.

Ice cubes don't fly to the moon because it's outside of their nature and neither do monkeys and yet both have been into outer space, that is, the space outside of the atmosphere. So when you calculate all this with relativity theory it makes sense to stay in bed of a morning, but there's no way in hell you can stay in bed forever when the coffee starts to boil and the fumes come around to get you going.

And this is called breaking the spell of inertia or the inertia theory of relativity and should not be confused with dunking donuts which is another theory altogether.

Advanced quantum theory is not for everyone but that doesn't mean you can't question everything, and maybe this is why salt was invented so a little bit could be thrown over the shoulder now and again.

To grow a theory from nothing does take some energy and lazing around in the sun of the tropics with nothing better to do helps a lot and in this respect having a lot of nothing to do grows theories in plenty so never let it be said nothing comes from nothing.

Random flying particles that sting and white elephants on the other hand take a bit of getting used to and this is why Confucius may have said: 'the ear is a marvellous barometer of impending probable events.'

Applying ice cubes to this probability effect is probably not why ice cubes would be taken into outer space, but then neither are monkeys there for that reason either; and the either, either or, theory can't be applied here to this and so not all is as it seems in this respect.

Which brings me to part five that will be told tomorrow; but it is said that there is no tomorrow or yesterday, only now...

INSOLUBLE SMILE

Last night the hadrons were scratching on the wallpaper trying to get in but this morning they were gone.

Many points south of the prevailing wind of this alley-alley-oh the king was singing into his megaphone star crossed sizzlers of never say die two moons removed and yet right on the button until he finished with: 'and we shall fight them in the bushes...' which was a little off the mark.

"Shut up you ole bugger," shouted Millie Bendy from out of her bedroom window, and like this you can see what I have to deal with living so far from home.

This is a criterion of my understanding that I have reached to date and like the universe continues to expand.

"Death-beds at the ready," announced the king who then promptly got into it and went to sleep forever.

Next morning he awoke with a hangover which just goes to show kings shouldn't sleep in death-beds while under the influence of any inebriating substances; which reminds me, I must update my journal.

Now where did I put it? I divine by this little rhyme something that begins with... yipped, yappety, yapped went the old hound dog in the aforethought a mentioned where a seemingly insoluble thing would be beneficial especially when it is seen too close, as most problems are.

So if a problem is an obstacle in the path, it makes sense to go around it rather than trying to remove it, study it or destroy it.

Philosophy is good for exercising the mind but when it's dark and you can't go on and you really don't feel like singing a song, when all looks hopeless and you're wondering what's the use of it all then philosophy may not be much help and maybe that's why all the old philosophers wore the inscrutable face or the insoluble smile.

When does a dream fade? When the door remains shut for too long.

And step out and fly because by then you'll have nothing to lose and maybe that's why everything has brought you to that, that can see this and know it for what it is, this that is the turning tide at the bottom of some deep dark well where you can fall no more, and looking up you see the light...

This covers the formidable exceptions in the relative loophole theory of redactive enlightenment.

A REGARDLESS DETERMINATION

If you can't get on with the small view of things then go past it and invite the big picture; and if that's not enough then go for the bigger one; you may find though that in the end you have to come all the back to yourself...

It's not that I don't believe in God, it's that I don't believe in what the system has told me about their version of God. I don't believe in the religions or anything they say and I'm letting all that nonsense fall away so that there's just me and my thirst to know. That's a clean place to be, a pure place and simple, no complications and the only ignorance there is my own and that I can deal with.

Clarity is clarity and sometimes it turns up as a waking moment when everything falls into place and all becomes clear.

In the moment of clarity you can see a long way and you know your choices and can choose them from a clear perspective and even if the choice is not easy you know what has to be done for the best.

But sometimes when a course is set and you're half way across the river it may be best to see it through or find yourself carrying around half a mind full of wondering, and did you do the right thing?

Sometimes a lesson has to be learned the hard way by doing it until it's done and then you'll know for sure; and never regret the decision you make, just carry it through until you come out the other side, maybe to know better next time; but it is possible that it will keep coming around until you go through it to learn it.

Life is not fixed, it flows and it can all change for the better in a moment and what you've always wanted can walk right into your life and change everything and the old circles will become a fading memory in time.

When a desire has been set in motion, everything falls into place to fulfil that, doors open, the right people come, money flows, signs abound to tell you you're on the way and all you have to do is keep stepping forward along that one path until you get to where your thoughts made beforehand.

When you get there you can reinforce what you wanted by thinking more on it or change the plan and make a new path to flow to; but if you find yourself without a clear feeling of direction maybe to the point of being directionless and wondering what you're doing where you are and all you can ask is why am I here then it's possible you've stepped off your path or been pulled onto a conflicting path just like particles in the particle accelerator.

This is the time to go deep and get in touch with yourself and then feel your way to what suits you best.

You'll know when you're there, your feelings will tell you, but if you still feel unsure then resolve the question in your mind by asking where it came from by following it back to its origin.

You may find it's an old asking that you've closed the door on or a gift you've not fully received that won't go away until you've accepted it; so it is always prudent to be careful of what you ask for.

If it's in your heart it's in your life and is there forever even if you're not around it anymore and time may muddy the memories but love never dies.

It is said that home is where the heart is, so lucky is the one who lives in that divine within for that one will always be at home wherever they are. True clarity is knowing that.

Sometimes understanding comes with a price that leaves you with a choice, that when you truly understand, you have no choice and wherever you are is only where you've come to, to understand that all your choices have brought you to where you are and that your direction in clarity will take you into a deeper knowing of your life purpose and that the thirst in your heart outweighs all other distractions and that your only choice is to follow it or forever be chasing the endless desires like a dog its tail until you end up with no more time and find that the endless desires have filled your life and you come to see them for what they really are, mere fading trinkets that pulled you this way and that while the real value of your life went unfound and by then it's too late, the last breath has come and your one precious life is done.

Fortunate indeed is the one who knows this in time, who realizes that the profound quality of life is inside and always has been and that the thirst is the ache to be fulfilled, to know that which can't be found anywhere else but in that heart of hearts.

Much has been said about this by the mystics so it is not new but if you see this and understand then let it be new for you and begin now.

'But begin what' I hear you say. Questions that fall out of the sky this way when the door of questions opens is the questing of the intellect to satisfy its curiosity.

How many failures does it take to crush the dream and beat you back into your shell?

What do we care about questions when the sky is bright and the road is free and we are forever young in our hearts?

A tree sings sweetest when it's in a shower.

All things change and some things change sooner than later so if there's one thing to be known it is that sooner or later whatever is known will be updated.

But when the dust lifts you faster than you can stand up, take a rain-check, you may have fallen over; and enjoy the raw-berries that are usually there at that time of falling.

A problem stops being a problem when it's accepted.

An old paradox called vaya con Dios sat by an empty dustbin crying tears that rolled down his face.

"What's wrong with you?" asked a naked girl wandering past.

"Nothing," he said.

"Really?" said the girl.

"Yes, really," he said.

"Then why are you crying?" she said.

"I don't know why but every time I see an empty dustbin I cry," he said with tears rolling down his face.

An observer peeking around the corner spied the naked girl and went: "Oow la la."

The oow la la bounced along the pavement until it got to the empty dustbin and with an alley-oop it jumped in the dustbin.

The naked girl immediately put a lid on the dustbin and that was that.

The oow la la could be heard making lewd noises from inside the dustbin.

"Not so empty now hey?" said the naked girl.

"No, it's not," said the paradox and got up and shuffled into the next story leaving the naked girl to do whatever she pleased in a story that would end here if it could.

Suddenly, as in bombs away and tally-ho there came the sound of an old paradigm clomping along the road searching for a way home.

Not to be outdone by this the acrobat of self consistency with its usual flair somersaulted right into this tale without a by your leave and then somersaulted right out again.

"We cannot be bought any other way but this," came the nervous voice of the epistemological pirate up through the drain cover as he cautiously put one foot in front of the other in the semi darkness of the tunnel he was explaining.

A tall philosophical ghost who had an aversion to spiders was drinking a merry juice with a straw in the shadows of all he could believe and was thinking that if he believed better in what he was thinking perhaps his thinking would make a better belief.

Round and round in circles he went in this while drinking his merry juice without any added sugar until one by one the cows came home and it was time for milking.

But never mind that right now we have protons on the move dancing with their neutrons in the ying-yang dance of interconnectedness where all things are connected in a soup constantly changing.

Spinning around in the circles of time and not time and colliding with the molecules of the attraction base in seeming impossible meetings called synchronicities the heart strings of the master magician being at work in all things, so it is said, if a little obscure at times.

A wavelength of precise mathematical certainty and probable distinction headed outward as the bow wave of consciousness from one relative being and collided with another certainty of probability and produced a quark that beat about the bush until it could stand upon its own quantum theory.

Everything is energy and no two things are the same even though they may be more alike than not in the statistical formulations of all that can be determined regardless of the theory, and so joining produced a mix of their molecular structure so alike and unlike the source as to be remarkable and in this way continuum is sustained.

When one subatomic particle meets another in the energetic playground of dancing energy the outer seeming is attracted from an inner urge and produces a random play that flows in a perfect game and becomes indelibly stamped upon a fading indestructible material substance not unlike dust.

All things being equal this has been supposed so until the next breakthrough in science that will show another thing to be true but perhaps in the eternal depths of the never ending and constantly changing the endless breakthroughs are but one small part of the constant creation of all we know that is and although certain constraints of understanding may be perceived as so on one level the black hole of uncertainty negates such on other levels where no one principal of thought will fit; which leads to the idea that what is currently known may only be one small stepping stone on a journey humanity has only just begun.

PICTURES OF THE REVOLUTION.

"There's a chemical imbalance in the grave and the ghosts are causing trouble," ran the headline of the morgue express.

"Finally some sense," said the bluebird of melody that had come down from nowhere into this place that is being made as fast as it's being made.

"So tell me, how do you grow this that can't tell what it is to be made more of?" asked the voluble insistence.

"Go through the door," said the echo of all that has gone before.

"Will it take me home?" asked the old question.

"Home is where the heart is so let that be your direction," said the old answer.

"But I came here from there to get to here; must I go there again to find my way back to here?" said the voice of the adventurer.

"You'll never know until you go through the door," said the door.

"Must I go around in circles forever then?" asked the adventurer to his ball and chain.

"Just patch me up and pick a direction and just go," said his old boots full of holes.

"One more coffee," said the empty cup in the traveller's paradise.

"Not so fast," said the noir book of blue disenchanted by the holy weight of life.

"Well then," said the mystic, "it is time to go home."

"Scratch this upon your wonder to make time for where we will live upon that hill of what we will burn there," said the burning picture hanging on the wall.

"We will dance around the roaring fire and be naked in amongst the shadows of our dancing and when we come down and go back our dreaming will be enhanced for having been there," said the naked girl who loved dancing.

"It's funny how one thing leads to another," said the old revolutionary revolving and taking his turn to howl in the night.

The uncertainty principal of consciousness leaves much to be desired sometimes in its quantitative tendencies to occur in the probabilities of its existence to determine laws that only the observer of them can correlate into qualitative forms albeit with a touch of existentialism thrown in for good measure to boot as holy sailors in our own depths to plumb and perceive as we can what is there.

But may believe if it happens couldn't suck on the straw of life any harder and was unwilling or unable to let it go and so became the straw sucking the life out of life until the aliens came and took her far away for debriefing.

Many underage developments were milling about and howling about this while the love of their lives were all at another party across town on the hoodoo tiles and letting rip.

Captain Barabbas of the Barabbas love boat out of Bangkok was nearby on a one way ticket and he was looking for his lost love that was lost never to be found ever again so he caught the next train to anywhere thinking that if she wasn't here then maybe she was there.

CQ calling; CQ calling, come in now your time is up," said CQ on her wireless out on the waves.

"Come bury your head here and wipe your fate on the understanding the fires are alight and burn our eyes with brilliance," said the naked girl who was still young to a party of spectators standing about doing nothing but waiting for the Hadron collider to finally work full time.

"Where are we and what are we doing here?" said the spectators to the Arabic bubblegum machine.

"Perhaps the fundamentalist approach through disputation raining reason upon the endless cycles of protruding thought is one accepted approach to the age old question of why we are here and yet after all this time no profound breakthrough has occurred to propound one answer for all. So common sense would dictate, for most sane people anyway that is, that another approach to this seemingly insoluble problem must be found," answered the Arabic bubblegum machine.

"And what's that when it's at home?" asked the Queen of Sheba.

"I like reality best after a couple of drinks," said the drunk.

"What reality?" said the reality to the drunk.

"I'm beaten-up and wasted by the way that it's all become the long story," said the drunk.

"Well if you live it you can tell it," said reality and flew away.

"Oh," said the drunk.

"I'm tired of being old and cold in the grave; how do I get out of here?" said the old cold body in the grave looking out at all the failures piled high to the horizon, which is when the old cold body began to see them as stepping stones to skip across.

But where would the courage come from in such crushing defeat to step out and fly?

Oh she's out in the rain down the bottom end of the thunder waiting for her ship to come in to see her through to the other side of her night, that long night she fell into down through all the long years of her life that were not so sweet.

"The words we beat ourselves up with are but suggestions of form, old fear thoughts that we allow to hang around and that now have become so hard to train," said the beauteous voice of reason to the rumbling train rumbling by with no explanation but its destination and the observations of its observers.

THE SLEEVE OF AN OLD MEMO

Most times we would have it one way as opposed to another but sometimes, as in the case of the missing enlightenment syndrome not all things immediately come about by wishing, and sometimes even things don't come about at all if the door remains closed to you even after repeatedly banging on it.

Some would say this is a good indication that you are knocking on the wrong door and that you should go somewhere else or come back at a more reasonable time of day, and sober, because not everyone wants a drunken clown banging on their door at four in the morning and heaven forbid you should bang on the wrong door and it opens to reveal the husband and angry karate black belter; you've had your chips then and no amount of quantum theory will save you.

This is called: being in the wrong place at the wrong time with nothing up your sleeve but an old memo never to drink again.

And now I would like to cover the electromagnetic theory of conducive conductors using the quantum theory baseline as our principal part of reference and I may even throw in some Chinese ying-tong theory to spice it up to boot as well.

The double slit experiment showed that consciously observing something can influence its outcome; but then they do say so many things on Saturday Street in the mornings.

They say there are things that have a one hundred and twenty parts per billion and I'm sure that's true; and there are times when a rusty chair is all you've got, but you still don't make use of it; and there's times you don't notice something because it's just quietly there and hardly doing anything but being; and when you've stripped something down until there's nothing left you then find you've gone too far; and Ledd Zepp said: 'what is and what should never be;' and sometimes that's all you need to say, there's no need to go and write a book about it; so shiver me timbers and rattle me bones if the expeditious magnetronic blame-wolves are afoot and watch out for the heat of an old steam engine if it passes you too close.

So take a memo Miss Jones to prove we are still here and send it to all the thoughts we've lost along the way, there's a good girl that you are Miss Jones, always bright and early wherever you go; oh those spear-heading electromagnetic waves you send ahead of you makes it all so easy; and rinky dinky do as they say in the pictures that I do believe they call the cinema these days if I'm not mistaken; and look here comes the menu of an old sea dog who never wakes up from his inebriations and's been asleep sleeping forever in the ying-tong and can't find his way home from anywhere because he doesn't know where he is and the way back is all lost now down through all the heat of time, and all the bridges have been burnt anyway.

So what are we having in the mustard this morning that we can call anything to name it so?

Circles in circles, round n round, come out backwards never to be found...

This principal of exception can be found most anywhere when the disparate whisperings are calling from the smallest measures of the mind, ghosts all, in the whirling, still attached from the past until the cords are cut and they are blessed and let to drift away to where they belong where the light can keep them in the memory of all the universal existence to date.

You can liken this to what has appeared momentarily for a short time on the human scale of things, but for its size exists for a considerable time and travels a long way and although it may appear to do very little, still, it has its purpose and without which no other thing could be.

We're talking very tiny particles here that make up the whole, and although they can't be detected without an extremely extreme machine they do come about to be seen after two protons have made contact seemingly violently as in a collision, but then all things are relative to the observer.

Quantum theory to the layman can seem dauntingly complicated with a whole talk-speak of its own and yet as with any discipline understanding comes from the practice and deliberate immersion in said such; so such said that this is so we have now covered the theory of existential ghosts in the machine.

AN INDIAN NEWT FULL OF SUGAR

Cultural identity could be likened to mass hysteria; all form being illusion and not real and any identity with it is a kind of hypnosis such as the old boys school of us and them and so identifying with this is a form of slavery.

Realization of this is a step towards freedom and self independence away from all forms of control.

In that independence the muddied waters clear and a profound stillness occurs and is a self awareness of being that can bring great joy; but for most the school of thought is that a fancy face can go further than being half in or half out and then some between the sheets of a full bloom; but like the poor proton in the machine of the experiment blasting off is not all it's cracked up to be before sundown with a camera in your hand when the quarks are around; and now by the miraculous decision to expand we shall presume to be innocent unless questioned guilty; this is a hot-seat you can't win so plead the fifth amendment and say nothing.

This is where we occur, where we appear and most often where we are and we shall clap for the winners who have won for no reason at all but that we've let them for they've come from the hard place and must win at all costs, and it's all right, we don't mind at all; do we Miss Jones? And if we sleep forever don't wake us, we are sound where we have fallen, peeling onions in the graveyard-shift of open expansion...

Cupid's lips but this is all a typhoid fever of beauties me bucko that is the all of it all, but you are all too late and far too soon in this awakening that can't go on without you and the heart to do so, and no one else can realize you but you.

And now the fifth dimension comes to ruin a perfectly good plan I have to run into the story lines of what I have seen to say more; but I am full of erudite like a prized Indian newt full of sugar that stands without, wasting to go within, and wishing, and I am ready to fall like a

Newtonian apple right out of the sky into those quantum arms and disappear into them forever; but how can I fall that far from here where I am on the ground with no falling beneath me?

If you look closely enough through the walls you can see patterns of energy. These patterns have edges that speak, but so far all they've said to me is that the Russians are coming with their machines that can read minds and even make you think things.

I keep asking these patterns if they can be more precise but they won't say any more, for now it seems, so I am left here to my own devices with half of an answer to all things.

OLD PARADIGMS

There are elements of old paradigms yet remaining with a lot of power and influence that would keep the intelligence and consciousness levels of the vast majority of humans considerably restricted to preserve the way of life of those at the top.

Things are changing so fast at an exponential rate a great awakening is going on and to the ones who have awakened enough nothing can be hidden and even though huge attempts have been made to dumb down the populace to keep them in their place as slave workers and consumers still, the universal consciousness level is rising which must be causing some consternation and fear to those very few who are in seemingly invincible and unassailable power over us, which makes them even more dangerous than they're ever been.

And so this thesis is my attempt to address this imbalance to explain quantum theory in such a way it can be understood for what it is by the ordinary person in the street, even though their attention spans have been disturbed like hot molecules to boil over at the least little thing.

Quantum theory, so named is but something that has been studied and known for thousands of years and the current modern theories are but explanations of what was already known long ago.

Such as the Buddhist theory that time and space are but constructs of the mind and that solid form is an illusion that is made up of whirling patterns with more space in between these whirling patterns than was previously thought; and also that everything is connected by an invisible force that can't be measured.

So if time, space and form are not real then what is?

So, do we keep on saying: 'pay unto Caesar what is Caesar's, or do we finally remove the ignorance that is sucking the lifeblood from humanity?'

Fear, as in fear of reprisal and condemnation may make you hesitate.

Condemnation from the religious authorities and reprisal from the state policy; between these two humanity has been kept repressed for centuries and maybe even far longer and who knows how much advancement we all could have made if greed and religious persecution had not come about.

If we come into being for but a moment as an illusion to experience uniquely an existence of form, and having seen to return, the form back into the dust leaving behind a small ripple that fades back into stillness, what then is our existence for?

Is it a self determined course we make for ourselves to eventually awaken and become enlightened or are we here for something else? And if we can't find it are we losers to be ridiculed; and if we follow one path for an age and still find no higher plane of existence beyond the mundane, have we failed?

For like this we are lost; and I hear your voices; and I would tend to agree with you if my bed wasn't burnt to a cinder long ago.

Generally speaking, what you see is what you get, but what you see is not all there is, so, on the one hand it might be said: what is the use of knowing this, and on the other hand you might say the use is in the knowing.

Brief is the love that comes to play sometimes, and quickly it slips away unnoticed; and sorry is the one who lets it go and tired to try again yet hoping for the heroism to feel where even in the aloneness we are not so alone; and a long way from heaven we are where we wear our feet so low and our splendour oven lower.

Small is the night without love and long are the hours to endure; and then to cry in the small hours of your love when you find your love is not so big; and all the Herculean endeavours have come to less than nothing, and less than that you feel so small in such a huge place where no physics can ever penetrate so deep to raise you up.

So the question is, do we carry on into the void, perhaps not even hoping anymore to find that heart, or do we turn back, or, if all seems lost then perhaps to just give up; and all this seen under a spectral kaleidoscope and is kind of like nailing down an onion on the edge of Jupiter with a piece of coloured crystal.

The gamma gallows theory leaves much to be desired.

NUMBERS OF THE DREAM

All is energy, but attempting to open a door without the right key can leave you feeling exhausted and so on any crossroads of redemption you will find many tired pilgrims reciting the door opening mantra without much initial success unless the door opens from inside, but I'll cover that in succeeding chapters.

Mantras are designed to take you on the circular path of least resistance and so are good for restoring energy.

The frequency of any energy signature opens all doors on that frequency; so when the frequency is assumed of the numbers of the dream wanted then seemingly by miraculous design, doors begin to open and the path becomes clear.

Mantras are the beginners guide to understanding vibration but it is a long road and is only part of the discipline of undoing the separation we all have of our own natural vibration.

There are quicker ways to enlightenment. One door you can go through is a back door from the use of psychedelics and or plant based hallucinogens.

Going through this door can be quite a ride but most often will leave you wanting more with a feeling that there's something more you've forgotten; eventually to realize it is an impermanent solution to a very real need.

The big door, or the door of doors is the door that can answer all questions, even those without answers; but this door is one you must find your way to, and when ready the master appears with the key.

This is a constant process of the student sitting down with the master and the thirst being filled with as much joy as it can hold.

In the field of energy it is much like the wandering quantum particle finding its way home and becoming part of the universal whole.

It is achieved, in part by falling through all the known layers to leave behind all theory and come upon the answer you've been looking for since you forgot it.

We can extrapolate upon this all we please but in the extrapolations is the separation.

Time being a constant construct and as such is not real and can only prolong the separation from the whole, so being in the now is a pretty good place to be and has instant benefits and comes with awareness and consciously being here and now.

All concepts of course are just that, and are miss-leading in their theory; much like a finger pointing at the moon is only the finger and not the moon.

A belief will only take you so far along the road of belief and unless knowing takes precedence from the experience that leads to knowing then belief can be a huge burden and is only part of the great illusion, the same as a theory is only a theory until it is proven beyond doubt and then it becomes indisputable fact, until someone delves deeper still and finds a more indisputable fact.

But all things known in the material world can be unknown and quantum physics is on the verge of this realization and for some there is a holding back with disbelief because such advanced quantum theory is beyond their limited belief systems and so denial becomes the state of affairs until it enters the mainstream of human collective consciousness to become the next set of beliefs to follow blindly much like the circular mantras designed for such.

Those enlightened and courageous individuals on the spearhead of new discovery are the ones making new ground and leaving behind the old paradigms, and so we should applaud them for their efforts and not hold them back.

THE EXCALIBUR MOMENT

One leaf after another of random thought falls into eternity forever like swimmers diving from the great height into the depths so far below and never to come up again. Ah the poor dears, where do they go so silently to disappear for so long?

But how long really is forever and never, and who knows but one day we'll pass them on our way home and say: 'Oow look, there's forever and never...wave.'

But like a trashy romance novel, most thoughts are not worth keeping and are best let go of to go their own way; so give me ten dollars and I'll tell you all about the Excalibur moment and how romance trashed the stone; and for twenty dollars I'll sing you a song from the borders where all the rusty words have gone and are lost wandering around looking for someplace else to go and the leaves of discarded thought fall out of the sky and are blown to dance in the wind that blows all things.

And so in that Excalibur moment when two thoughts collide the energy produced spins more from out of nowhere to hop skip and jump onwards, forever onwards pushing the edge of knowledge and all known things even further; and this is why hippies used to say: 'far out man,' in the cutting edge of their movement which was an explanation for all things in the making.

There was once a time when these thoughts were called madness for they were so far outside the concepts of the rational mind they could not be understood and were reasoned away as an illness to be treated with medications and psycho analysis and named as mad.

There was great fear in being mad, in the western culture, for mad ones were locked away, sometimes for life, and if you weren't mad when you went in you became so in those confines.

So come take my hand for comfort and dispel the old fears and move onwards in love where our thoughts rain from our soul to cover our paths as rose petals of pure energy to light our way.

ESCAPE CLAUSE

The longitudinal latitude of Faraday's escape clause states that Newton's law cannot be bent more than the speed of light or levered less than the rebound theory of relativity; which is to say that there are more loquacious events happening in this universe than can be counted; and so because of this there is a regardless theory doing the rounds that says that intelligence on the planet is in general less than recommended for what is needed to fly to Mars and has come about through lack of enlightening education or dumming down by insidious means, or both.

Needless to say, this total amount of ignorance must weigh heavy indeed upon those in charge, which is to say that if they were supplanted by actual wise souls then humans would progress in leaps and bound upwards.

The unequal balance that is caused by the greed of a few is a cause for concern for many and must be redressed soon for the good of all of us, for it is causing destruction on a planetary scale that is not sustainable.

It does seem to be that the only allowable points of reference for role models that the population has are the leaders, presidents and prime ministers and that all demonstrations are to be controlled, banned or subversified and any role model outside the system such as JFK, Lennon, Gandhi, King and others who become too powerful suddenly die so that there is no rallying cry for humanity to organize a turning point to end the corruption.

So it is incumbent upon us all as individuals to oppose the old paradigms of darkness wherever they are seen and like the sea, to continually wash it clean until it is no more.

Of course, if we all rose up at the same time we could overthrow the usurpers but then it is possible that such anarchy would not in the short term be beneficial and may cause great harm, and so perhaps enlightenment must come to the individual first and then it can spread to the world.

The marshalling yards of extremism are overflowing with anger, so hang onto your hat if you decide to go there.

So that's enough gossip for now.

GETTING BIGGER

And now if you can all stand up without falling over backwards we shall proceed to the observatory to observe the new worm-hole made by the Hadron Collider that is extremely close to our planet.

Yes the hole is getting bigger. Yes it is swallowing anything into it that comes near and yes the ghosts are making noises down below, and the aliens have phoned home, and yes we are working on the problem, and no we don't know enough to know how it appeared or how to turn it off, and yes all the aliens have gone to Mars and yes all the politicians went with them and no they can't return for at least two years and yes that's how long we've got to make a new world.

ADDENDUM OF PERCEPTION

So what is the answer beyond the allegorical statements?

There is energy and by focused intent such energy can be gathered and harnessed for good and healing.

It can also be used to overcome inertial resistance in such a way the now deliberates the Tao of all you are in the moment and great insight can be experienced.

Such a state resists external energies by its acceptance, much like the sand of the beach resists the power of the waves by allowing until the force is spent, and in this way calm can be achieved.

Great insight is achieved from the flow that arises when conscious thought is deliberated from subconscious intuition.

Such clarity when exercised brings a depth of understanding that raises mundane thought above itself to a level where that which was hard to perceive can then be known.

Doors of perception open, time slows down, the organic mechanism of creation is released and the profound is seen in all things.

The nature of truth is a vibration found within yourself and comes from a knowing of your deepest self which is connected to all things.

It is known that one who knows their self is hard to lie to, for a lie is a twisted vibration and no matter what the face may show, a lie cannot be hidden from the high vibration of truth.

This is why the aliens of humanity will do all they can to keep the population dumbed down and distracted in consumerism, fear, division and chemical inertia...

THE SPIRIT WORLD

In the spirit world, everyday events are seen through a different light and so can be coloured by the need to see beyond the veils, but once entered the turn of events is most often beyond conscious control, for the novice and so a guide is desirable and in most cases necessary.

Those who enter the spirit world without a guide, more often than not become lost and try to turn back, but once the door has been gone through it is no longer able to be seen.

Becoming lost in the spirit world, even for those with expeditious fortitude is no matter to be taken lightly for your energies can be preyed upon to the extent you may not have enough strength left to find the way out and so will become a ghost and wander lost forever.

Most animals can enter and leave the spirit world at will and so having your animal guide with you as companion and helper is sensible and wise.

It can be a shadowy place for some and for others like a great desert where nothing is as it appears.

To enter the spirit world means you must be able to see the energies, that is, the energy signature that is given off by all things. Reading this energy will indicate to you its nature and intent and so you can avoid the pitfalls.

Hunger can appear as a snake. Lethargy can appear as a dark bottomless pit and excitement as circling crows waiting to pounce, so remaining calm puts less energy out that can come back to bite you.

If you find words coming at you from a benevolent entity you may find that the words were already in you and will be known and you can trust them.

But beware the twisted knots of the strange, from other entities that may not be your friend that will try to convince you to follow a path that has no heart.

There are those who return who seem changed somehow, not the same person who went in and it is sometimes wondered of these ones were taken over, supplanted by a spirit that wanted out of the spirit world, while their real soul walks in the mist forever lost.

A spider web in the spirit realms can be a web of energetic patterns leading inwards to the centre carrying knowledge, and so this can signify finding the centre of your power where your true knowing can be found, and from there your path of heart is known and you can pursue it if you so wish.

Your animal totem can help you with this if you pay attention and are aware enough of what may be shown; wisdom is in quietly seeing what is revealing itself to you, until you know without a doubt.

Great pain can break open the veils of the spirit world and take you, if you let it to the space where you are completely alone; and great good fortune has the one then if the source comes to find you; but no matter what, say nothing with your mind for this will send the source away, for the mind is an interloper here and has no place when the great one reveals your true heart to you.

Many have said: 'all things are within you,' and this is so, but what can be seen if the tempest of the mind is not silenced in the circles of whirling?

The spirit world is but another realm we can experience to gain understanding, and as we are all connected, when enough know a thing it will spread to the wider consciousness and be available to be known to all.

More and more are awakening and finding that there is a great beauty to be known and although this has been hidden for so long, the veils are dissipation and it is no longer so hidden, and so there is hope for us all as more and more of us turn to see the truth/beauty that has been there all along and that is the birthright of us all and not just for a few elite who would lord it over us.

SOUL STONE RIVER

In the soul stone river where one goes to be immersed in the waters of understanding, the enchantments are re-set and clarity comes; all doors are unlocked and a great freedom descends to wash clean all illusion and scatter the bones of sorrow as dust upon the wind.

All great souls know of this and have said it in their own way to anyone who would listen.

But how do you say what can't be said?

The heart has secret ways to speak that the intellect can't comprehend and in this way Knowledge is passed on from one to another in its most profound and simple form.

A baby breathes through its heart all things and in the breathing knows it is as one with all things and so sleeps in the peace of a tranquillity only babies know.

PARTICLE OBSERVATION

So what have we learned so far?

We know that the basic stuff of the universe is made up of small particles and that these small particles move so fast that they appear solid when in fact there is a vast space in between them, so vast is this space that if the particles were to stop moving the universe would wink out of sight instantly.

It is also known that these subatomic particles are themselves made up of smaller particles and so on down smaller and smaller.

It has been theorized that the universe is 99.999 empty space and that if all the particles that make up all that is of the seven billion humans on Earth were to be put in one place with no space in between them it would amount to the size of a sugar cube.

So seen in this way nothing exists except what is seen in the mind of the observer, or the dreamer if you like who dreams into existence all that he sees, and we, as the participants of the dreamer can also dream into being our world as we would have it, or remain asleep and dream the ready-made dreams of others.

All is energy and we have the power to influence it by our thought and heart waves.

The new thinking of science is coming to the conclusion that there is no conclusion and that the universe is not so empty but that it is filled with something that holds the universe in place.

The Eastern mystics have known for thousands of years and so in this way are thousands of years more advanced, or further along than western science, which has been held back by the old paradigms of religions and political thinking/beliefs that has sought to stay in power by influencing the thinking of everyone by mass fear, disinformation, control and subversion of the energies to such a low state that no upwards movement was possible for the vast majority until now with the advent of the internet whereby information is available to all.

Yet even now the dark powers exert influence to hold us all back and like the matrix where we are all plugged in and feeding the dark energies it is incumbent upon us all to break free, to reclaim our own power and use it wisely for the benefit of all.

In this way the more that do unplug from feeding the darkness the more it will diminish their power until the turning point comes and the tide turns and light finally comes to all, and the darkness withers away and is gone.

THE NATURAL ORDER

The true test of a warrior of light comes in the vulnerable moments when the heart is open and the armour is off.

The perceived subversive energy comes to disturb in such a way it tags its shadow within you to pull you to fight, to react, and if acted upon rashly then the limited mind of panic thoughts takes over and you've already lost.

The advanced warrior understands this and sees it for what it is and becomes calm, no thought, until the moment comes to diffuse the opposing energy and return the disturbance to its natural order.

But some energies are like a volcano and boil over and keep doing so until they have vented all of their power out, and until they do so, calm is upset and patience is needed.

A wise man of course, moves away from an exploding mountain until it is prudent to return.

PART TWO

The Quest for reality.

We are what we think. All that we are arises with our thoughts. With our thoughts we make the world.

Gautama Buddha.

So did Buddha become Buddha by thinking it?

ZUM

Reality can be said to be all that is left when everything you think you know has been stripped away to leave the grace of what you are to shine.

If everyone knew of this the world would change instantly. The great illusion is in not knowing this, even though there are many who speak of this, there are few that listen and know.

Perception is subjective; whatever you subjectively perceive is of your world; so you could say there are seven billion odd subjective realities running around all with a different perception of their existence; but is reality different for all of them, the reality being all that which is not a dream?

Religions and politics have forced a conscripted reality made up by them on to us all and anyone who says otherwise is consigned to the hell fires down below, and the unlucky ones, the real heretics get to work in the boiler room stoking the boilers, which is the hottest place; but conscription aside, illusion is illusion and can never be reality no matter how much pain and death and lies are exerted and served up as truth and the one way.

Those who know this have already seen it and those who don't know don't want to see it or have not yet woken up.

Whose job or responsibility is it to inform the public of injustice, wrong-doing, malpractice and etc? When a lie is seen for what it is, shouldn't it be exposed?

When the aliens suppress the truth then it is time to stand up and say it is not so, and that perhaps they should not be in office.

Many feel this way and many more are waking up. Those who do and say nothing even when they know something is not right are condoning the conceit.

Those who speak against it take a chance on being ridiculed, hated, ostracised, and condemned and in some cases even put in prison.

So it seems that freedom of speech is only what the powers that be allow and as more people do speak out and protest then freedom to speak out against them is being curtailed more and more to the point it has become a crime to say the system is seriously flawed.

It could be seen that the unrest all over the world is not terrorism but more and more people standing up against them.

Who is this 'them' that makes the 'us'?

It can only be those that somehow keep getting back into power, the ones who are supposed to govern and rule justly for one and all and yet are only out for themselves and their masters, those ones behind the scenes who insidiously use their wealth to influence events, the outcome of which benefits no one but themselves and their lackeys.

But as I've said, fighting this lends only to more fighting. The answer must lie within each individual to be aware enough not to be fooled by deception and to live as best they can from a heart based reality.

STRAY FEELINGS OF REALITY

There are many forms of reality that can absorb your attention until you become quite lost in them so that they appear to be real and unassailable, aiding and abetting to keep your consciousness snared until it seems there is no other option but to believe in them.

A bad trip will end sooner or later and so long as you keep that in mind it can be endured and survived, and time will take care of the rest.

The mind is a great usurper of power and without the heart's wisdom can run amuck. A wise choice is to allow the heart's intuition to guide you.

Sometimes this will come as a stray feeling to not go any further along the path you are on, to turn back and find another way, or maybe even to just get back into bed and sleep that excerpt of reality away that you find yourself in at that time.

Some realities can change in a moment and upset your world to the point you may find yourself without a solid foundation to stand on; great pain can wash away a lifetime's reality leaving you debilitated and finding no room for any more thought based theories to such an extent you become withdrawn; but pain can be an enabler and if followed to its source great enlightenment can be found.

This reality supersedes all others and some have said it is the reality of realities for there is none deeper or further in: the source of all joy, and when in that presence there is nothing but that joy and the profound reality where no illusion can enter.

As human beings we have minds and it is our own mind that can intrude to pull us away if we listen to it and then back we'll go to whatever our life has brought us to; and yet, being touched by that friend, by such profound kindness, by such, will leave us changed forever and no other reality will suffice to be called real, whereby showing everything else to be illusion, and when this is known no belief can shake that knowing.

IS IT TRUE?

It is perhaps true that no one else's reality can be known, for language cannot take the place of experience; that expression is personal and is subjectively portrayed from our own unique soul.

This makes every single one of us a unique being, one of a kind, never to be seen again and as such, priceless in existence, and so, no life can be bought or lost at any price and those who put a price on life have cheapened themselves; such low energy could benefit from seeing the light.

Is it true that when you wake up the world wakes up?

One seventh billionths...

If the world is asleep, can you wake up?

I wonder sometimes who will read this and if it will be of any benefit to them; and if knowing the aliens are not good for us all will do me any good in the long run.

Voicing my concerns is my right, whether anyone will take any notice of this is not my concern.

"What about the worms that ate the biscuits?" groaned Titus Groan.

An exhausted chuckle on the hugeway became disoriented at this turn of events and decided to go lie down for a while.

And I, of course will not be letting every heckler raise my heckles thank you very much. Time to bring on the dancing girls. Full steam ahead there captain.

NUMBER OF THE LU-LU

The mind quests forth away from the answer.

The art deconists have other ideas on this in the fifth year of the bee where the lulus are taking over.

"We shall all become wandering visionaries," they say; and upon this they build their foundations to fall over again and again.

"Who are we to fall this way?" they ask, falling that way to be.

The answer is ginger tea of course to shrink the brain back to where it was before; oh such a rich miracle in Rosemary's bar of herbs; and passion juice is the answer for the nodding dog's head in the car window.

Well that's what you get for coming here with a spoon; bring a shovel next time.

But now the number of the lulu is blowing dustbinious askings again so if you really want to talk in Greek I'm sure I can find a few geometric patterns to throw back at you; which brings to mind Stan and Laurel's pie fight; but never mind that now.

Ice rice and the average lifespan or two quantified ovenously might make you call out: huvun as they say in Cornwall before the English invaded and took their language away; and now we shall never speak Cornish again, but more to the point, ovenous statistics aside for a moment I still have my brain power and the Hadron Collider has not so far taken it.

Cotton stitches, I almost forgot, the combination to this puzzle is: $a/b=(a=b)/a=p"phi"=1.61803...$ Which I hope you'll agree with me, even though long overdue is never too late.

ATTENTION SPANS

My attention has been aroused recently at how soon life mirrors my thoughts back at me.

This is an observation with some concern, for if my thoughts are making things then I should be careful of what I think.

The mercenary in me raises the question as to whether I can direct my thoughts along the lines of attracting abundance in some form to better my life situation; but if how it is, is from how I am then the change must come from inside at an intrinsic lever.

But thought can only take me so far and then comes the time when I have to trust that when I step off into the unknown along my life path, I will be taken care of and be looked after.

And that's the answer to the bucket full of grace question.

50 SKINNY MISHELS

50 skinny mishells all dimping and motley were making mishamy in the bishellooshy when suddenly out of the blue sky jumped a ZidderZie. Oow.

I do believe intelligence breaks down in the face of so much inknown, perhaps as a release valve, for if you carry on into too much void, and alone, there comes a time when you become spaced out, with too many parts of you in too many places at once, and all those parts, questing for some kind of reality to hang onto, but in the void there is no foundation to hang onto but yourself, not the ego self, or the intellectual self, or any other of the constructs of the mind, but the real self of the heart; and so simplicity is proffered in those moments when congestion comes to bottleneck all forward momentum; and of course, having some empty pages left on the passport for the galactic officials to place their stamp to facilitate your movement from one place to another.

This is another form of control on the free passage of Earth's children from one zone to another that they have mapped off as their own.

The Earth belongs to everyone and no one.

The aliens have much to answer for.

But oftentimes it has to be said: we are only here but for a moment, and after we've gone, we never come back again, at least not in the same way; so this is it for all it's worth, this is our time.

It's not the clock that tells you your time, but the beat of your heart and every breath that comes; and holding the flower of consciousness open to experience, you can expand into all that is.

And now we shall listen to the thirsty fish in their ocean: glug, glug, glug, glug.

THE SQUARE PERHAPS OF UNCERTAINTY

...and then you find love has gone, and left a desert behind that you wander in beseeching its return.

You call out: 'come back,' but it has turned its back on you and left you alone; and even if it does come back for a while it would be but a shadow of what once was, and then going again it would leave behind the pain of the world and a wreck of a miserable worm that can find no comfort.

Any who come near at this time, maybe out of compassion would see a depth of sorrow that eats upon itself and absorbs like a black hole any good that comes near.

The hurt of separation is the greatest pain a human can go through, and some do not survive it.

Only time can close this rent in the soul, or direct experience of the source; indeed, one moment with the source takes away all the pain and reveals joy there, unparalleled, and wise becomes the one from such contact.

But few get this far; the raw wound of their pain is mostly too much to bear and so salves are applied such as drugs, booze, sex and even suicide in an attempt to alleviate the pain and make it go away, but this pain is too close to home and no drug can get behind it enough to take it away, just maybe block it for a time which can lead to addiction and then that's another problem to be dealt with.

Eventually the pain does go away and the tears dry and you begin to pull yourself back together.

It is said that it takes about two years for the cords to unravel and fall away, but perhaps they never do and that the pain is just buried enough for life to carry on with some kind of normalcy.

For some, healing never takes place and the heart remains broken forever and what is seen outside is but the fragile shell of the broken inner self.

But most do heal and get past it and rise up the stronger for it and then go on to love once again.

So if you find yourself on a row boat in the middle of the ocean one day and it's not raining then think yourself lucky.

FROM THE JAPANESE

Half a laugh later and the hypnetic veils were lifted by the slippery hand of doom turning on the spit of will it rain or not; and then the morning sun came out.

What can we see where we see so much?

Here is a sandwich for when you get hungry and some orange juice for when you are thirsty; and here is a love letter for when you need to know someone loves you; and here is a suitcase full of money to buy what you want.

So it has finally come to this that is not that as if it mattered; but who am I when I'm at home to shift the brick from one foot to another to expose it for what it is and then to move on along the wall taking out one brick at a time until it falls over; but seriously, what would I do with a suitcase full of money?

The queen will now ride in her carriage in the universality of it all and I must live in the fall, the prophesy has foretold it so, with the wet rain falling sideways into my eyes.

"Awesome," said the snake in the hot sun.

The road eventually comes back upon itself, the beginning the end, the end the beginning; and as it is in, so it is out; and not a penny-whiskered farthing can say otherwise in truth, but we know they do.

"Is this why we must fight them on the beaches?" asked the hard-luck story.

"No, you must fight them where they press your buttons," said the reply.

"You laugh in disbelief; do you think you'll live forever? Time goes by in the blink of an eye. Stop blinking you there at the back, pay attention, it's almost coffee time."

"Pass the rocket and hold the used comments and a bag of bliss on the side but no sugar," I said enthusiastically on my third beer.

"Do you want it in Japanese?" asked the bally-ho.

"Are they naked?" I enquired.

"They will be dressed sir," said the bally-ho smiling his inscrutable smile at me.

"Well ok, as it comes then," I said feeling like I might have missed something.

The evening went swimmingly with not a mosquito or cockroach in sight and afterwards I walked home in the surreal moonlight and felt pretty good all things considered, but I still haven't figured out what to spend a suitcase full of money on.

Maybe it's because I'm a slow starter or maybe it's because I'm afraid to let it all in, I mean, money is so vulgar; when you buy something it's a statement and reveals your inner workings, but I guess when you want something badly enough money is an easy way to get it.

When you're content in a discontented way that you can't explain and you have a whole lot of money and you find you have to begin spending it soon or lose it, that's a kind of quandary that puts you in a spin and then the whole world is for sale and the clocks are all ticking and you find yourself doing things you wouldn't normally do; like shopping.

You walk into the big store that sells everything and you walk around and stop occasionally to ask the price, out of habit, and then you end up in the cafe drinking coffee and feeling exhausted without having bought a thing and begin planning the easiest way to get back home; and when you get there you flop down on the bed, your pockets still bulging with cash and you groan with the weight of it all.

Of course, when the wife comes home all my problems will be taken care of and I can just lay back and watch old French and Japanese movies while she goes out spending; such a civilized way of getting rid of money.

But I can't tell my wife; not yet, maybe not at all.

There are times when I feel I should get out there somewhere and look around for something, and occasionally I do find myself out there and looking, a little bit, and feeling lost, but it's a thing of a thing that I find myself doing and my imagination fills in the blanks of the whatever to the nth degree.

And then I find myself turning off and dreaming, and then a poke in the ribs brings me back and agreeing to whatever she's saying: 'yes dear, of course dear, your wish is my command,' which usually pleases her and I can get back to my daydreaming.

But a daydream once disturbed is a hard thing to get back into, even when you walk back into the exact place you were disturbed from, it has moved on and you're left without the connections to carry it on, and then you find yourself half in and half out of somewhere and looking around hopelessly and feeling: what is this and how did I ever allow myself to get here like this in such a way.

Most days it is like this, unsticking myself from all the things that are so sticky, or feeling out of synch like a bicycle wheel rolling down hill with no one chasing it to become lost in the weeds forever where it will finally fall.

And there's nothing to blame, not even the full moon where the hallucinations run wild and carry you through the rain of a crazy night to the crashing ocean where you heave in the sea's roar and spit in the eye of doom until morning's light finds you curled up in your bed taking comfort from the radio and the fact you've survived it, again.

And when you wake up it's all a day in a day and you know there's nothing to compete for, and mysticism doesn't even come close to what you're feeling, or not feeling, so you delegate all the meetings to love and know it's alright, there's coffee and little things that tell you you're still you even though it's so hard to put your finger on what's real.

My wife came home and brought with her a bottle of wine and I wanted to let her so much to take me away with her but she wasn't going anywhere, so we sat close and drank the wine and laughed at nothing.

And then Danny came on his motorcycle and heaved a crate of beer in and told us of his travels of far away, and we listened so intently and thought how daring and free he was.

I curled up for an afternoon nap and drifted off to Dylan's: I shall be released.

I woke to the rain beating on the tin roof and Jasmine, my wife's sister, who'd come along while I was dozing and was making soup in the kitchen. I think she'd heard that Danny was here.

And then Jean came who was Jasmine's rival for Danny's affections and went through into the big room where it all was going on, or so it seemed.

The next thing that anyone knew was Danny and Jean were in the back room; I guess she'd taken Danny away with her, and later when Danny left on his motorcycle he took Jean with him and I never saw them again.

Friends come and go like this until one day they've all gone, and then one day I find I'm gone and time has erased the road back, and I find myself in rustic restaurants talking Japanese badly to anyone who will listen to me, my pockets full of money, and dreaming.

When the boat came in and I was safely onboard I felt glad, no more decisions to make for awhile except what to have for dinner and which movie to watch. You could even say I was happy.

I'd buried the suitcase in the garden wrapped in cellophane; and now I was going around the world on a big ship, everything taken care of.

Just before the ship left I thought better of it and got off and went home to my loving wife and kids; what had I been thinking of?

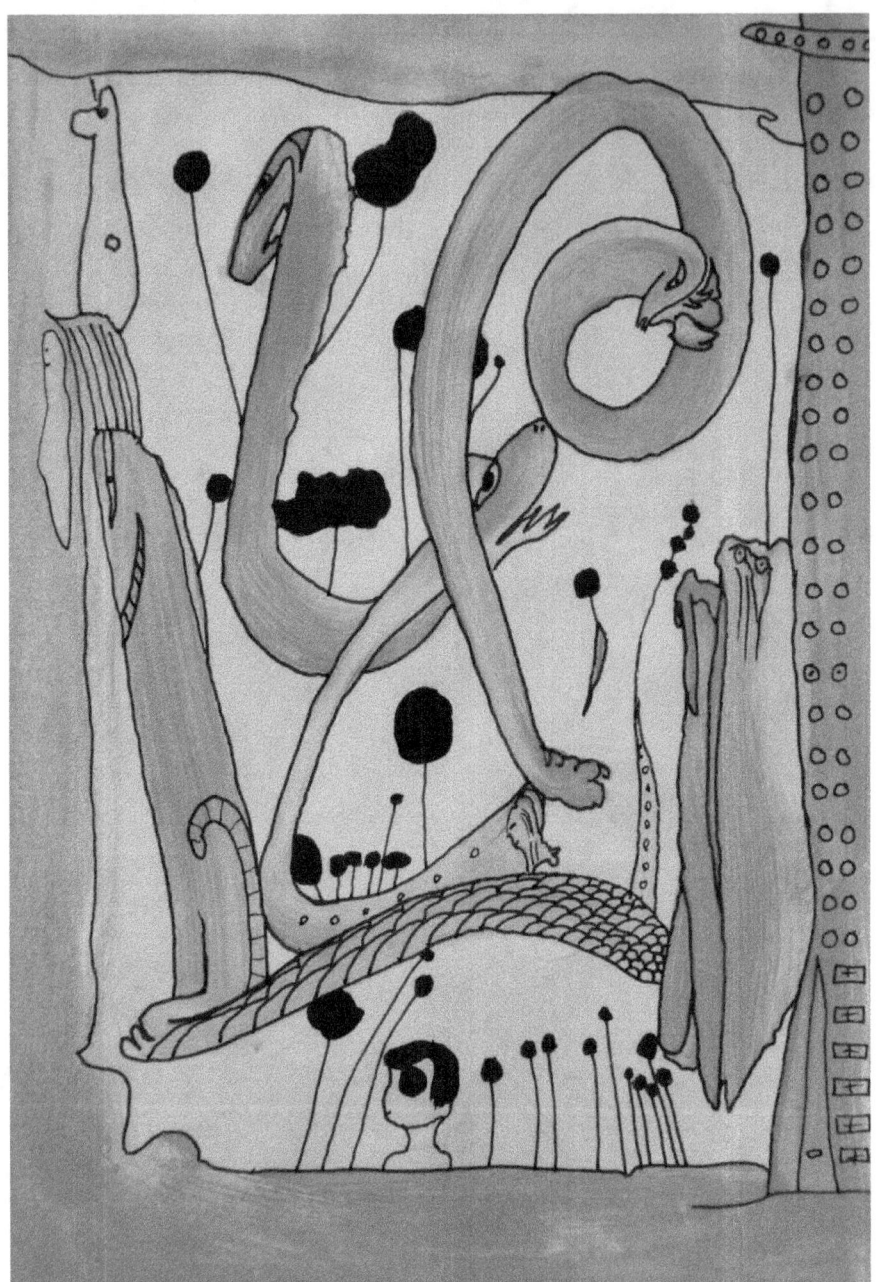

PICKS-A-PICKLE

Success or failure; does it really matter? Hmm; we are whatever we're feeling.

And now I shall reveal the secret of the universe in a five span hand clap called Mu. Did you get it? Never mind, see me later after sundown and bring a bottle of wine.

The feeling machine burns with an incandescent brightness from inside until only fear of the unknown is left and then that too is consumed and transmuted into energy that is far beyond the old paradigms.

Beings of light understand this and listen, but those stuck in their minds fear it and call it madness and would suppress it.

The celebration continues, the masks falling away to reveal the true beauty from inside.

Many come to know, to ask for the truth to be revealed; and for those who are sincere nothing is hidden.

So, if we're all still on the page and not gone dancing it means more must be said...prepare to be eaten...

WHAT IS REALITY?

You know that time when the shilling expires in the shilling meter and the electric pops off and everything goes pitch black and you find yourself there, half way through brushing your teeth in the basement where the tap is; so you carry on with your teeth routine by feel and then you're reaching for the towel beside the sink somewhere when there's a noise behind you.

You know there's no one there, the room was empty when you came down, just the flagstone floor, the door you came through and the three other doors at the other end of the room.

The other doors were closed, but you know there's nothing and no one behind them because only you live in the house and you've been through the doors and know what is there: on the left is the coal cellar where the coal comes from a round manhole in the pavement above and is too small for anyone to squeeze down.

The middle door opens into a rubbish space that is full; and the third door opens into where the gas cooker is and a couple of shelves with saucepans on.

There's one other door which goes out the back where the toilet is down the end of the yard, but you've locked that door.

So you're standing there in the pitch black and waiting for the sound to come again so you can place it in your mind to know what it is, and as you're listening you're questing around with your senses to pinpoint where it is and feeling vulnerable and finding you're suddenly terrified yet holding it down because you know that if you give into it and scream you're done for.

Slowly you edge towards the open door to the stairs up and put your foot on the first tread knowing that you're in the lair of this thing now; and there's no help, but even if there was it couldn't get to you in time.

As silently as you can you creep onto the stairs and wonder when it is going to come for you as you slam the door and hold the latch down.

But you can't stay there like that all night long holding the latch to stop it getting through the door and anyway, it is probably too strong and will tear the door open and get you.

It's a long way up the stairs to the front room where the candle is and the monster at your back all the way, but you know you've got to try.

As you dash for the top you lose it and panic and begin clawing your way to safety, the thing behind you breathing down your neck.

Somehow you make it to the top not knowing which moment will be your last and fly through the door into the living room desperately searching for the candle and the matches to light it.

There's still a little glow from the fire, the coals eerily sending their last red light into the room before turning to ash; just enough light for your fumbling fingers to grasp the slippery candle and light the short dark wick.

The candle's lit but there's hardly enough light from the tiny knob of flame flickering to die as you as you stare at it, eyes large as saucers and holding your breath so you don't blow it out.

You pray to god with all your heart to keep the flame alive and not let it go out.

The coals settle in the fire and make you jump.

Slowly you turn towards the open door to the basement but dread moving any closer to it to close the door for you know something's hiding there around the corner and ready to reach out and grab you and pull you down, back into the pitch black of down there and you screaming all the way.

The candle seems to be holding its own against the darkness that is ready to encroach and envelope you and squeeze the last small drop of sanity out of your terrified mind.

You clutch the candle in both hands and look at the last door to go through to safety.

You close your hand around the handle, your eyes sore and red from not blinking for so long, and press down on the latch with your thumb and hope it doesn't squeak and make the bat flap wing monster tear into the room filling it with its hideous shadow and devil's claws and more you are too terrified to even think of.

The door opens and creaks but you slip through the too small crack and slam the door behind you, and then there's the long climb up the final set of stairs to the top where the bedrooms are where you can climb into your bed and lie there unmoving until the morning light comes.

As you ascend, the candle flickers and spots of candle grease fall to the treads of the stairs and splatter onto the thick greasy globs of the ten thousand previous nights you've had to do this same journey and you think maybe it would be better to just let the monster have you than to go through this every night, just turn around and face it; but that way lies madness...

FOR WHEN YOU BECOME A MASTER GURU IN THE MOONLIGHT.

Many flaming arrows come and talk about starlight and moon-beam but when I fizz them up like champagne the cork stays well in and the passion I would see is so tied down the balloon cannot reach to the skies.

In the importunate moments of this discovery the chains are rattled and the conversation turns to something safe.

We are not ready comes the deepest thoughts hidden behind walls of steel.

But we know about this so let's talk about that...

I have seen the chink in their defences and so tickle them there, unobtrusively while they're not looking and suddenly I find all eyes are upon me, for the heart knows and would listen; but the secret is not something you can say; it is sometimes seen as a spark in the eyes that shoots across and ignites another, and then the search begins for that one to ignite you forever.

But say no more while the heart revels in its discovery of joy; if they want to know more then they will come to enquire in some way and there must be an opening to that...

OVERCOMING MING

And now without further ado we will begin the big-road walk to go somewhere, but we know not where...

Without a plan it could be a long walk but we are eager to get somewhere; so we will walk until we arrive, or find ourselves at the beginning, which is where we usually begin from and end up, and who knows but that maybe this time it will be different.

So, is this the definition of hope, or folly? Or perhaps it is something else...

When you are lost what other choice is there but to stay where you are until found or walk in circles; and this is why overcoming Ming is so important, and if you see the blue-haired lady in the loophole you may be getting somewhere so have your portfolio of questions at the ready, just don't ask any of them outright, in fact, don't say anything at all, just follow to see where she takes you.

P.S. watch out for the black-holes along the way.

Oh, and some doors that are opened can't be closed again so the question comes from out of nowhere: why would one try to close them again?

Tired is the moon that looks down on this; the worm shrivels-up and the dog bites its own tail as the ship comes about full of new thoughts while the old ones die away a death that doesn't bring them back again.

"Headache on the horizon," is the shout.

"Steer away," shouts the captain.

The doors are left far behind; tomorrow for sure we'll all go home.

Do not speak now, there's nothing more to say.

Boom-de-la-boom went the out of the out past the psst where the ning yang was scattered and round the round where the round was round and then up the down where the inside is out where someone said: we're nowhere near when we're so far away.

Boom-de-la-boom of the din that's in of the Texas gin that makes us spin.

And now for the bent pancake; oow la-la.

So I went out for a pizza and cut my eye tooth on baby talk with the girls who spoke in German but I had to turn off from that pretty fast without understanding a thing they said.

And then they sang a song and got up on the table top dancing and moving like girls do.

I smoked a cigarette and stared at the cactus on the table willing it to grow but I don't think we were on the same wavelength because it didn't do anything, not a bloody thing.

I remembered my coffee all of a sudden on the table and so had a sip. I like coffee but it makes me hot, but I always have my room to escape back to so I can cool down.

I was desperately desperate in a quiet way and wanted to fill this yearning I had that was bothering me.

One or two of the girls smiled at me in my tale of one in the growing dark and beer time was rapidly approaching, I could feel it in my bones where the wind would blow and who knows what would happen, but perhaps I was picking it up from them, the girls, and if only I could get past the watershed of my mind the mystic could enthral me again and all would blow beatifically to fill the table in the now that was rising.

It seemed like it was all going on, if a little relaxed on my part, and yet, I wasn't growing anything on the tablecloth of bright colours; it was a still life of objects, some subject of small pieces that were just there.

The girls' movements were candy out of the corner of my eyes and were trying to hypnotise me, so I kept my head down a little bit to not be so distracted.

I know the best part of the night was here in the hope and aspirations of something more to come later, some letting go in a heavenly way that would surpass the sleepy heads falling all around me.

Spraying my body with mosquito repellent I surrounded my table with the cloying smell and hoped it would do the trick.

The girls creaking boat came and they were called, and hugs and calls saw them off on their adventure.

And now twilight was promising to open all the doors of the night to be able to sail through to my heart's content.

And then I was in heaven with my pizza on the 9th parallel of the emptiness that was knocking to come into the space that was now and blessed.

The magic was growing wings and I still had my coffee to finish; but maybe I should have said just coffee for it's all a gift and nothing really belongs to me.

Bubbles came and I found myself surrounded and I hid there in joy for awhile. Will they judge me for this and apply a spider tax?

Let them try and my laugh will drive them away back to their holes where they quantify their percentages to be applied to anyone they can apply it to, to acquire money that isn't theirs to take.

Those suits of calculation live in a world not of this one, and I don't subscribe to it, and when they come as one day they must do, when they find me what can they do, I have nothing to lose, but let's not let fear enter here where the dance is danced in the feeling of joy.

We are Einstein; we are Ghandi and Lennon and all of our heroes that have gone and those that are still here; we are so much more than we settle for where our burning burns us away into the dance of dances.

But I'm Monday maybe on Thursday and time is change; my yearning grows as I carry it around and I look through it all I see; but nothing can fill it.

The rain will come soon to wash it all away.

MARGINAL ERROR

I got some boozes in the snoozes and so hiccupped and sneezed.

We shall go soon where the going is to be gone to and bless me father for I have sinned I'm sure and so give me the bread of heaven in this sinner's need to get back out there again to do it all over with all my sins washed away. Oh give me a Hail Mary at least to blow away the dust and I promise with all my heart I won't sin ever again until the next time comes.

But oh no I cannot pay for my pockets are empty so please put it on the tab to be paid later when I come back down.

Bed time drives the chickens into their nests in the trees but I am hungry for more where the empty is filled in the bottomless pit where I fall over into for no reason to hit my head on the iron bed and hurt a little bit so some anger comes and I feel it to get said of the pain of the dark eyes in this place without bliss where too many cigarettes call for another beer and busting all the politically correct, the outsider staring at the past that's gone yet show so many faces.

In the boom-boom now I find I am alone and in the wrong place to rise above it, so crash and burn or find a place that is turning.

The new is all around but so still and the morning is so far away from me, too far to see from here.

But I am here and ask: why am I here?

Soon I will howl and look for the container of all my discontent as I fall into the bushes with the spinning head and the yearning for more.

Time booms the night, but here there is no music, just the machines that never stop their onslaught.

Where am I?

I come to the conclusion that no one is where I am at to come slow with what they breathe and in that breathing give a clue of where this is at.

The machines masquerade at me bamboozlements but it's not music and now I feel my yearning will drive me away from here to out there where I am, to out there where I am a stranger.

But I must be careful of what I ask for, a marginal error could leave me stranded somewhere I would have to spend years getting back from.

It is late and quiet and falling here and there to finally make it back to myself. Oh but the urges, the urges. What marginal error brought me here and not there where all I want is available to me, while here I am to waste behind the mountain of all achievement smitten to be bitten.

But I am afraid of myself to show my delight that burns out before the end to leave some awkwardness and wondering what happened and not sure to go there again, and so alone and hanging on for what?

A wind blows the answer to me and stirs up everything I thought I was; it has a message, something indivisible. I look from a place of not much to be filled with more; but I'm not listening; maybe I should pull down the wall that stops me hearing.

I've given up on reading other people's books, my own are good enough to read if only I could see them again, for once written they blow away with some unreadable wind that loses them somewhere in some desert they come from; maybe one day they will come back and haunt me in my rustic prayers for redemption; but even that door is fading as I hang on to some grounding to even be able to say this.

If I am to be blown away then let it be in a heavenly way, let the gargoyle thunder and ghoulish lightning become the rainbow to the other side where that beckoning receives me in open arms and I am relieved of the desert's journey that has been so long and arduous a trail to follow and be lost in until all direction becomes the same and no tread to any points of the compass is worth the following and only that star I began with to lead me on even though hope has been left far behind and the tired steps to the oasis falter and no answering howl in the barrenness of it all answers.

If I grew up again from a seed would I ever set out again on such a path if I knew what it would bring? And that I must burn my very life on its journey and leave behind all I can't carry with me, and under such a hot sun all things are released as a burden until all that is left is the very air I breathe and my pumping heart that moment to moment carries me onwards.

SHOULD I STOP DRINKING COFFEE?

Well then, and so it seems and so it is but what a strange room I find myself in.

It is said I make my own life by what I think, so what is it I see and feel and understand that I am? Is my consciousness so low that I cause my own downfall every day? How can I change that for the better?

Should I meditate more? Stop asking for help? If I move on, again, will I find a better life situation? Should I stop drinking coffee?

My mind is powerful and I know there is little I can't make with it, and I have, down through all the long decades of my existence; but so much time spent in all the making, and after all, I find I have one suitcase of stuff and a homeless situation that some say is freedom and admire it and wish they too could do such; but what do they really know of this, and yet, perhaps for them they could do a better job of it.

Ah, all the striving and effort; all the places and people that have come and all blown away now in the wind that blows all away eventually.

I don't know how much time I have and I would use it for the best and for that joy inside and what wouldn't I give for that?

My lips move and mumble some kind of prayer but I ask: did I fall off the boat? Or am I being looked after?

I feel so much to give up, and perhaps I have.

I would surrender now to the divine in my heart; I would let it all go, and be taken care of, or not as will be in such a life with a yearning that feels so strong for something.

Is it time? Am I ready? Am I already there? Can I move the mountain for the joy? Perhaps it will be moved for me.

So many tears have fallen where my heart has felt so broken so many times, yet I've got up somehow and walked on.

Why do I cry so much in this desert that waits for the rain, and waiting I fall and rise.

I find I am attracted to the quiet ones, to the soft hearts, to the beauty, to the smile of love. My heart aches for such. But so many distractions come and go, and passing leave behind something of their nature and such noise is bothersome.

Did I say there's wind blowing to follow my heart? But to where? So I look for directions and as soon as I see it I will go and never look back.

Perhaps I came to this place to heal my wounds and in the healing find my way, and in the finding to know who I am, and what; for what am I if I am not this body; and my mind is only something I look through, it is not me.

What am I? Am I the sunrise that sets my soul on fire? Am I the love in all things? When the full moon in the big sky comes to fill my spirit with its beauty and heartbreak, am I that which sees it through the eyes of wonder? And when my soul ignites with such yearning can I not penetrate all the veils to find my beloved?

Let me smile from my heart and set the world on fire with my love, the lover in love.

But look at these words that try to say what can't be said. My beloved is beside me all the time.

THE CYCLES OF THIRST AND THE FUNNY ONION IN ITS LAYERS

From the little pip comes the idea, but it's a small thing in amongst the huge beliefs and terraforming thoughts that rake the illusions into formulated plans and lines of direction I become drawn into and am surrounded by and then become until I've forgotten who I ever was so that every day I have to tell myself who and what you am.

So every day I look in the mirror to see some face I recognize as me and yet sometimes I wonder, who is this that looks back at me; and as I look deeper I begin to look and see something I hadn't noticed before: little lines appearing at the sides of my eyes. It's a small thing but it makes me realize I'm not as immortal as I used to feel.

I cry out: "Honey, come quick, and bring the magnifying glass."

So my sweetheart rushes in thinking the worst and hands me the big magnifying glass and we both stare through it.

"Look," I say, "can you see them?"

"See what?" she asks.

"Those lines around my eyes," I say, staring intently at them.

She looks deeply and then looks at me like I'm daft and gives me a swipe on the head and goes back to her make-up muttering that men are just too strange before coffee in the morning.

And then the kids, who have been waiting for us to finish and have seen this exchange between us, rush in and want to see it too.

"There's nothing to see," I say, but they clamour on, so I pick them up and let them look, and they stare most intently to find what I saw, but all they see are three big faces.

The wife passes the door and looks in and shakes her head and walks on to the kitchen calling out that it is time for breakfast.

The kids look at me and see only their dad. My number one daughter whispers in my ear: "It's gone now daddy, come and have breakfast."

I say: "ok" and they jump down taking with them the magnifying glass and run in glee to find things to look at.

I tell myself it's ok and finish up and go to eat with my family and yet I wonder what is this that's me that changes imperceptibly? Who am I?

Who else will notice the changes and maybe look at me funny?

All day long I look in people's eyes to see if they are looking back at me in some strange way. They do look back, and some smile, but it's my observing that changes what I observe and because of that I'll never know if anyone's looking at me strangely.

Many things come at me and bother me until my thoughts become a noisy engine I have to switch off somehow so I throw myself into my work designing TVs that the company can use to look at who is watching them.

This is the first layer of the funny onion that I have uncovered but I don't know it yet; and the cycles of thirst are in me now and with this, come choices I'd rather not have.

It's funny how a few lines in your face can bring restlessness like this and keep me awake in my bed when normally I would be sound asleep by now.

"Am I still what I once was or am I changing into something else?" This is what I think in the dark night as everyone else sleeps.

Do I slip this into the hand of the gods or let it go; either way I might get some sleep around here sometime soon I say quietly so as not to wake the wife snoring beside me and blissfully unaware of this turning of thoughts that rages in my mind and just won't let me go.

Yes, yes I was a baby boomer, but what's that got to do with it all this time of night? And I really do have to go to work soon and so please let me get to sleep.

And then I wonder, who is it that won't let me sleep?

I'm going mad talking to myself. I almost lean over to wake the wife to tell her all of my problems but this is my concern and I'll deal with it.

I suddenly realize that I'm growing old and the huge thirst comes over me again to know who I really am and what I'm doing here on this planet with my little life I use to run around consuming things and signing away my freedoms to work for money to perpetuate my enslavement to the system I've fallen into somehow without even realizing it.

It's all too much trying to figure it out and even though the chains bind me so tight that there seems no way out, tomorrow for sure I'll figure a way out of all this, but for now I must sleep.

Hours later my mind has churned the night away and the dawn imperceptibly lightens the curtains and creeps in through the cracks.

In all my turnings I have made my way to the stone's resistance that I cannot find a way around; and with that comes anger, and sadness. Anger that some stone of helplessness has been placed in my path to hinder me and maybe even stop me from moving past it; and sadness that the beautiful life I have is no more than a fairy tale that will end at midnight if I cannot find the key to all this.

Why do I feel so helpless in the face of such seemingly permanent impermanence? And who are they that would bar my way? What right do they have to do that? If I have committed a crime to be so imprisoned then I would like to face my accusers and tell me to my face whose law I have broken to be so un-free in my own life.

But I do wonder if looking in the mirror is good for me and that perhaps the next time I shall see something else even more disconcerting.

I climb out of bed carefully so as not to wake my slumbering wife and go get an early coffee and turn on the TV and see the latest atrocities.

Every channel I turn to shows some inhumane vengeance and how much the world owes to the banks and how I have to tighten my belt in the new world order austerity measures.

And then I see the super rich on their yachts laughing it up; and the government in their fancy suits giving themselves huge pay rises, and I wonder why they are so immune to all the austerity while the rest of us have to pay so heavily?

And then I think that perhaps they own the banks and are in total power and that maybe they are the ones who have put the stone in my path; and I begin to feel like a small pawn in a huge system with no way out.

I hear stirrings from upstairs and realize I can't speak of this now, so I put on a good face as best I can and turn on the lights and begin the morning's rituals.

I've opened a box that can't be closed. Who can I talk to that wouldn't think me mad; so best to keep this to myself for now until I know what to do. For now I will go about my day as normal

ON A SIGNPOST BURIED UNDER THE MOON

It wasn't that I was trying to forget anything but rather to change the thinking so I could feel better, but whatever happened, I wasn't going to jump off anything higher than the pavement, and as it happened, later, the pavement was something I fell over quite a few times, I even fell over a wall; but that's what happens when you do the dervish whirl with the warrior in you.

The next day I woke up with bruises all over so after putting antiseptic cream on them I decided to give up whirling, and of course, I spent some time in the dog house when the wife saw the bruises, but she knows me well by now so I didn't have too much explaining to do.

Later, after breakfast when the warrior came we spent the day walking in the warm rain and tried not to touch too much on the fire of any heated emotions that would burn us up.

But as the sun fell out of the sky at the end of the day I opened up and fell inside and then on a signpost that was buried under the moon I found a note from the thief that said: 'I will come for you when you least expect me.'

This threw me off my stride momentarily, but then I thought: 'yes, there are many turns that go unnoticed to trip you.'

I decided to become as aware as I could, just in case.

"You cannot win or lose in this battle; for there is no battle, there is only you," said the warrior, speaking up after a long silence.

"I'm a husband with a family, and I don't fight," I said. "You go your way fighting every battle and ready to fight at a moment's notice, but I prefer to stay closer to home." I said this quietly hoping to keep the calm.

"You avoid life and hide behind your woman and lose the opportunities that come to you in waiting and indecision," said the warrior, his shadow towering over mine.

I could see he was ready for a fight in any way he could provoke me so I became quiet and said no more.

"I will give you this, my known, that nothing is known but what passes in the moments," he said, looking at me for a comment, but when I said nothing he carried on.

"I can show you a thousand ways to die, but there is only one way to live and that's in the now."

"We all die sometime," I said and then wished I hadn't.

"The art of war is not in winning or losing but putting up a good fight and dying in honour. It is being strong in who you are and without letting fear sway you, give your all; and in those insurmountable odds against you your death will come easy or hard, it will not matter either way, it is how well you have lived that counts," he said enthusiastically.

"I don't understand," I said.

"Remember this then, that when you come to face your fear, the battle will be with yourself, and in this there is no winning or losing, for if you fight you lose, and if you don't fight you lose. And if you win it will be but your ego grown huge."

"What do I do then?" I asked, feeling this was beyond me.

"Die."

"I don't want to die."

"If you don't you may never really live."

"I'm afraid."

"When the time comes you must choose: to die and in so doing you live, or to run away again and never know."

"My wife might not like it."

"It's not her battle, it's yours."

With all this talk of war and dying I'd lost awareness of the thief and his promise to come when I was least aware and so in suddenly remembering I looked around to see if I could catch a glimpse of him, perhaps hiding in the bushes by the river bank I was walking along, a circular route I made when I needed to be alone to think.

"What are you looking for?" asked the warrior looking about him too, his hand on his sword.

"The thief," I said.

"Ah, the thief," exclaimed the warrior in delight and relaxing when he knew there was no danger close. "When he comes you won't see him, you will only know he has been when you discover how much he has taken from you."

"Is there any way to guard against him," I asked, feeling lost all of a sudden in such a precarious world where fighting, death and thievery could come at any moment to take all I had.

"The thief hides in you and will come in your unguarded moments to help himself to that which you would not lose the most, and after, when you wake up you will find him dancing in glee for having tricked you again so easily," said the warrior with no hesitation as if he knew well the ways of the thief.

"Perhaps I can hide from him," I said, wimpishly.

"You must wake up," said the warrior, his eyes bright with the fire of life coursing through him. "You have been hiding all your life in something and the thief has found you every time. Where you hide from yourself is where he finds you."

"But I'm not hiding, look I am here," I said, feeling I was on the verge of an argument.

"We have had this conversation many times," said the warrior, beginning to fade, "and each time you have declared you are the impermanence of the flesh you wear which is but a fancy marionette moved by your flashing thoughts and emotions that come and go."

"Then what am I?" I asked, unsure I really wanted to know the answer to that for I had a suspicion there was a lot to lose in the knowing.

"Only you can know that," he said, and then he was gone leaving me puzzled and not a little disconcerted by all his talk of things I only half understood.

I found myself close to home and yawning so I took myself off to bed. Tomorrow maybe I will get to the bottom of all this, but for now I'd had enough.

FALLING DOWN

I chiselled my way out of my dreams drenched and exhausted from fighting the thief all night long with the warrior's sword.

After a long cold shower I decided to be alone for the day and just do ordinary things like read a book and get a large coffee from a cafe and to sit and not think anything at all. And if it rained I would watch it falling and if the wind blew I would flow away with it in my dreaming.

Ready to leave the house I was just picking up my umbrella when knocking came at the door that I had to answer. It was the kind of sound that touched a nice place inside of you and made you want more of; so when I opened the door and let in the scent of patchouli, and Jasmine, who was my wife's sister following close behind, I was surprised.

"Hello," I said, my senses overflowing with the smell of her patchouli.

"Good," she said, in her no nonsense way, "I see you're ready."

"Ready for what?" I asked, wondering if I had the energy.

"To escort me to the temple of course," she said, her hands on her hips, much like my wife does.

'Oh-oh,' I thought, 'there's no saying no when she's like this.'

"Is it a mystery why we're going there?" I asked, giving her a half smile and hoping she would fill in the other half. But no such luck, she was on a mission and I was her prisoner.

Oh how helpless one is to lift a finger no braver than a squeak.

'Where's the warrior when I need him,' I thought but said out loud:

"How would you have me in this thing?"

She looked at me with a sideways squint and then grabbed my arm and we were off.

So before I could get my umbrella up and no sooner than we had stepped on the pavement the rains poured down and we were drenched.

Her clothes clung to her body and no imagination was needed to know the form under them.

"Eyes front soldier," she commanded.

I was having trouble with the umbrella, and in a sudden gust of wind it collapsed, so I dumped it in a bin and we poured on into the rainstorm.

A closed door goes a long way until it's opened and then out falls yesterday to begin again what was locked away forever. So perhaps the trick is to never open any doors that have the feeling emanating from them: do not open this door.

But perhaps when you fall down the pavement coalhole into the cellar and its pitch black and there's only one door and you really want to get out of there because it's creepy and you're beginning to get frightened and your cries are absorbed by the thick stone walls and you're on an escort mission to get somewhere then I guess it's the perfect time to suit-up in your hero suit and become the hero and to take stock of what you have.

Anyway, what could possibly happen if the door was opened?

Suddenly you find yourself on a mission and you can feel the awesome power building in you; no coal shed can defeat you.

You are now in invincible mode and superman has nothing on you; so you do some exercises to pump up your muscles and suddenly the door opens and light seeps in to show granny with her coal bucket staring wide eyed at the maniac in her coal shed looking like hulk and black from the coal dust.

What would Buddha do in this situation?

So thinking fast I drop to my knees and hands in prayer I bow and say: "oh great wise one, thank you for coming to save me."

"Get out of my coal shed," she screamed and stared around for a stick to beat the intruder who looked like all the bad man she'd ever seen all rolled up into one horrible apparition.

So I scramble past her on my knees with her screaming out and beating me with the broom she'd found.

'Got to get out of this crazy place,' I thought and headed towards the door at the other end of the cellar that was lit with a flickering candle.

I scrambled through the door and up the greasy stairs to freedom, but at the top was a man in a dirty suit leaning back against the wall and grooming his teeth with a long sharp knife.

As I ease past him warily he asked me if he could interest me in some cigarettes.

I'm still soaked from the rain and the coal dust is running down my face in seams as I make for the front door calling out: "later," to him, and then I'm out on the street to find the wind has died down and the rain has stopped.

Jasmine stands there in the hot sun, steaming and waiting for me.

"Will you please stop doing that," she says sternly with her hands on her hips and her hair streaming wet and her clothes still clinging to her from our drenching.

"If only people would close their coal holes," I said, feeling splendid after having escaped from another dark place.

We decided to go get cleaned up and dry and as we walked back almost everyone stopped to stare at us which was not too surprising seeing the state we were in.

Jasmine said not a word and I couldn't blame here really, she does try with me, for her sister's sake, but more often than not we come back too soon after some mishap and then I don't see here again for some time.

But what can I say? If it didn't suddenly rain and if people didn't leave holes in the pavement for me to fall into then I do believe things would be different and we'd get to where we were going.

Never mind, it's almost time for coffee.

QUANTUM OF CONSCIOUSNESS

"Which part of your existence do you not understand?" asked the warrior from out of the blue.

"All of it," I said without looking around to see where he was; he would appear when he wanted.

I was staring down at the pavement as I walked along and being mindful of holes that might suddenly appear and swallow me up.

"What belief pronounces that in you that you choose to believe it so?" he said, appearing at my side.

"Slip me into this where I breathe for all I'm worth, I would breathe that and not lose a moment where I can't be found," I mumbled, still thinking of black holes.

"You're mumbling," he said as we trudged along together.

I glanced at him and wondered how such a huge being could fit into a body like any other.

"I don't know what I believe anymore," I said, and realized how empty that sounded.

"Buried deep with you are your core beliefs, you may not be aware of them or remember how they got there, but they drive your life. They are the foundations of your thinking; your reactions to all you encounter are motivated by them; change your beliefs and you change your life," he said as if I knew what he was talking about.

"Look, it's like this," he went on with perhaps more patience than I deserved, "in the beginning you inherited your belief system from those closest to you: your parents or guardians, the ones who raised you, the schools you went to, your friends and all those who had some influence over you; and you, like a sponge absorbed them all, and then they became yours, and then you lived by them.

The beliefs grew stronger in you until you became them; and now it seems you can't differentiate between your beliefs or reality."

"But..." I began to say.

"Listen," he said almost impatiently at my interrupting him, "somewhere along the line you inputted something that is causing a conflict in you belief system. You are going to have to dig as deep as needs be to find what this is and then resolve it."

With this said he was gone leaving me confounded as usual and thinking that the hole may be bigger than the sum of its parts.

I abruptly turned around and went home to see if my computer had any answers to the problem.

Rushing in through the back door that is always left open I shouted out to see if anyone was at home, but I had the place to myself, so I set to work.

First question: does one belief lead to another?

Second question: what do the waves sound like when you're not on them?

It was a hole within a hole peeling back the layers of an onion that opened up more questions with a thirst to know that would be astounding if only the answers would come in a timely fashion and that I could understand them.

The computer was so slow turning on that I lay down on the couch to wait for it to warm up and fell asleep or what seemed to be a sleep for I was aware and yet not fully conscious and I found myself on the borderline in the quantum of consciousness where all answers were available to me.

I was in myself and seeing from me and as I looked I saw many things, but just as I began to understand something profound my wife and kids came home and I was suddenly awake, and although I wanted to go back there I found I was living in the moment without any explanation and that my family were my lifeblood and just being with them was the most important thing in my life, and all I wanted was to express how much love I felt for them.

THE MYSTIC MONKEY NIGHTLIFE

A subterranean idea half in and out of what I couldn't remove was taking a short circuit through the backyard of all I believed and was being swept away in the massage of the spreading dawn and losing all support for the beliefs that couldn't be believed.

These beliefs that were coming from somewhere most often and causing the thinking that if we're all connected then this may be so even withstanding the perpendicular in a remarkable way.

But later in a strange place I came to that had no doors the echo of my heart preceding me, and though it was loud it was in those confines I found I was inside where I wanted to be.

And then the dark came in its daily occurrence so I went out to eat and wander amongst the towers and trees.

"Have you come for the nightlife?" asked the mystic monkey to which I had no return.

My neighbour in his three thousand horse power wheelchair passed me at the speed of sound and going like the clappers, so I didn't wave; maybe I'd catch up to him later when his battery wore down.

The time came to turn inside out which wasn't hard and then I had no more ideas and found it a good place to be.

And so without floundering I began my search for a good place to sing from in the lover's embrace.

From an equation that was constant in the dreaming of many, the idiot and the fool attached themselves to my shadow and followed me in and out of here and there until they became lost among the bright lights of those that came to dance.

Momentarily feeling light I skipped in this until the inside turned around to find me walking along the many paths of abbreviations who were making it once and for all and would never see the end until they were old, not knowing that you never grow old, that it's only the outer appearance that changes.

The feeling of a familiar exchange began to spread from the place I'd discarded it to and although I tried not to listen to it, it influenced me enough to turn my back on what I should have heard.

And so began the process of finding what I hadn't lost until the spreading dawn came again in another suburban idea I couldn't remove.

It was here I realized I was repeating patterns and wondered which set of beliefs brought it on to keep me stuck and going around in circles.

As with so many mornings that came and went like this I found myself worn out and close to the river of no return.

My neighbour came around the corner of depleted expectations still in his by now well worn wheelchair and gave me a tired wave. I returned his wave with one of my own and wondered if he'd enjoyed the jazz and my nightly performance in the crooner's lounge.

But not all the gravy in the world can move a mountain of soup further than it will flow unless it comes to the end of all things and falls over the edge into the eternity below.

And then I thought that this is why sleep was invented so you can go somewhere to forget about it all.

www.ingramcontent.com/pod-product-compliance
Lightning Source LLC
Chambersburg PA
CBHW070812180526
45168CB00002B/586